日本酒
SAKE

东京艺术之旅
TOKYO ARTRIP

〔日〕美术出版社书籍编辑部　编著

朴惠　译

中信出版集团｜北京

前言

旅行指南系列《东京艺术之旅》，每册一个主题，带你从日本文化、艺术及设计的角度畅游东京。每期均有不同的艺术之旅顾问登场，本册以日本酒为主题。第一章中我们将迎来清酒酒吧经营者、俄罗斯品酒师德米特里·布拉赫；清酒造诣颇深的料理研究家 Yoshiro 将在第二章出现；向世界传播日本酒文化的日本酒信息馆馆长今田周三以及旅居东京的和歌山酒藏第七代传人长谷川聪子分别在第三与第四章中为大家担任日本酒讲解顾问。他们将各自从不同角度介绍品尝购买清酒的绝佳去处。想要进一步了解清酒的你，一定不能错过。

Introduction

TOKYO ARTRIP is a series of guidebooks about Tokyo. Each playful walking spot introduced in each edition is selected from the perspective of Japanese culture, art & design. Several ARTRIP ADVISERS appear in each edition. In this book on sake, the following 4 ARTRIP ADVISERS show you around the city: Dmitry Bulakh, a kikisake-shi (Master of Sake) from Russia who also runs a sake bar (PART 1), Yoshiro Takahashi, a cooking specialist who is well versed in sake (PART 2), Shuzo Imada, the director at Japan Sake and Shochu Information Center who promotes sake to the world (PART 3), and Satoko Hasegawa, a seventh-generation sake brewer from Wakayama who currently lives in Tokyo (PART 4). These sake specialists suggest bars and liquor stores chosen from their own unique viewpoint. We hope this book will give you deeper knowledge about sake and help you to enjoy it more.

"藏家"店内（详见第98页）
This photo shows the interior of KURAYA (p.98).

目录

第一章 —— 8
品日本酒：享受新时代清酒

- ❶ 藏葡 ········· 10
- ❷ 麦酒庵 ········· 14
- ❸ twelv. ········· 18
- ❹ 希纺庵 ········· 26
- ❺ SOREGASHI ········· 30
- ❻ 银座君嶋屋 ········· 34
- ❼ 采 ········· 36
- ❽ TSUNEMATSU 久藏商店 ········· 38
- ❾ 未来日本酒店 DAIKANYAMA ········· 40

第二章 —— 44
品日本酒：那些魅力主厨经营的小店

- ❿ 青 ········· 46
- ⓫ 松 ········· 52
- ⓬ SEKI 亭 ········· 54
- ⓭ 旬菜 SUGAYA ········· 56
- ⓮ 四季饭晴间 ········· 58

第三章 —— 64
学习日本酒的相关知识

- ⓯ 日本酒信息馆 ········· 66
- 什么是清酒 ········· 69
- 清酒的种类 ········· 70
- 解读清酒标签 ········· 82
- 挑选清酒酒器 ········· 84

第四章 —— 96
买日本酒：那些品位独到的酒水臻选店

- ⓰ 藏家 ········· 98
- ⓱ 小山商店 ········· 102
- ⓲ 升本 ········· 104
- ⓳ 内藤商店 ········· 108
- ⓴ STAND BAR MARU ········· 112
- ㉑ 合羽桥 酒之 SANWA ········· 114
- ㉒ KURAND SAKE MARKET ········· 116
- ㉓ 和歌山纪州馆 ········· 118

- 清酒小知识 1 ········· 42
- 涩谷的酒鬼横丁 ········· 61
- 小泽酒造 ········· 86
- 清酒小知识 2 ········· 94
- 乡土馆名录 ········· 120
- 小百科 ········· 123

符号说明　时 营业时间　电 电话　休 休息日　址 地址　交 交通路线　网 网址

※ 如无特殊说明，料理全部为不含税价格，酒类全部为含税价格。
※ 书中内容基于 2018 年 10 月信息。

CONTENTS

PART_1 — 8

DRINK: NEW STYLE SAKE

1. KURABUU — 11
2. BAKUSHUAN — 15
3. twelv. — 19
4. KIBOUAN — 27
5. SOREGASHI — 31
6. GINZA KIMIJIMAYA — 35
7. SAI — 37
8. TSUNEMATSU KYUZO SHOTEN — 39
9. MIRAINIHONSHUTEN DAIKANYAMA — 41

PART_2 — 44

DRINK: RESTAURANTS WITH WELCOMING OWNER-CHEFS

10. AO — 47
11. MATSU — 53
12. SEKITEI — 55
13. Shunsai SUGAYA — 57
14. SHIKIGOHAN HAREMA — 59

PART_3 — 64

LEARNING ABOUT SAKE

15. JSS INFORMATION CENTER — 67
    - WHAT IS SAKE? — 69
    - TYPES OF SAKE — 70
    - HOW TO READ SAKE LABELS — 82
    - CHOOSING A SAKE SET — 84

PART_4 — 96

PURCHASE: SAKE SHOPS WITH UNIQUE SELECTIONS

16. KURAYA — 99
17. KOYAMASHOTEN — 103
18. MASUMOTO — 105
19. NAITOSHOTEN — 109
20. STAND BAR MARU — 113
21. KAPPABASHI SANWA 720 — 115
22. KURAND SAKE MARKET — 117
23. WAKAYAMA KISHUKAN — 118

- TRIVIA ON SAKE_1 — 43
- SHIBUYA NONBEI YOKOCHO — 61
- OZAWA SAKE BREWERY — 87
- TRIVIA ON SAKE_2 — 95
- LIST OF LOCAL SPECIALTY SHOPS — 120
- GLOSSARY — 123

Icon Description (H) Hours of Operation (T) Telephone Number (C) Closed Days (Ad) Address
(Ac) Access (U) URL   ※All food prices are fax-exclusive and liquor prices are fax-inclusive, otherwise stated.
※All information contained in this book are as of October, 2018.

第一章

品日本酒：
享受新时代清酒

DRINK:
NEW STYLE SAKE

第一章

清酒简直就是"米酿的葡萄酒",
那些用全新形式传递清酒魅力的店铺

俄罗斯人德米特里·布拉赫小时候曾因父亲工作的关系,在日本住过一段时间,大学时又在日本做过一年交换留学生。为日本饮食文化所倾倒的他在毕业后来到日本,开始从事饮食工作。对于清酒,他清晰地记得"第一次喝时我就被惊艳了,心想这简直就是米酿的葡萄酒"!如今,德米特里已取得品酒师资格,还曾在品酒大赛中取得第三名的好成绩。他为我们带来的全都是时下东京最有个性的品酒好去处。无论是为了便于品味酒香而选用高脚杯盛酒,还是在酒与料理的搭配上十分讲究,每家店无一例外地都在以一片赤诚之心面对清酒,以便更好地将清酒的美好传递给客人。

Sake is indeed "Rice Wine"
New Wave Bars That Bring Out the Best of Sake

Dmitry Bulakh, having lived in Japan as a child due to his father's work, returned to Japan to study at a university for a year as an exchange student. During his stay, he acquainted himself with the art of Japanese food culture. Since graduating from university, he has been in a food culture venture in Japan. "The first time I sipped delicious sake, I was struck by its flavor. That was a rice wine worthy of the name!" confesses Dmitry. He is an excellent kikisake-shi, who won the third prize in the 4th World Kikisake-shi Competition. He recommends bars that are all rare and unique to contemporary Tokyo, including one that serves sake in wine glasses for more aromas to be enjoyed and another that is very particular about pairing food and sake. Each of the owners are fascinating as they strive to learn everything about sake to present the best of it to their patrons.

ARTRIP ADVISER
艺术之旅顾问

德米特里·布拉赫
Dmitry Bulakh

品酒师。出生于莫斯科。幼年旅行及交换留学的两次访日经历让他彻底为日本饮食文化所倾倒。在担任清酒酒吧"twelv."店主之余,还在经营售卖日本茶等食材的网店"极东杂货"。

A kikisake-shi born in Moscow. After living in Japan twice, as a child and as an exchange student, he became fascinated by Japanese food culture. He manages a sake bar "twelv." and operates Far East Grocery, a web shop handling Japanese teas and other gourmet goods.

❶ 藏葡（筑地）

由千叶酒商"Imadeya"运营的现代居酒屋，主打日本酒，地处筑地警察署后侧，一个近来因个性餐厅聚集而颇受关注的地区。店内的所有酒品全部经店员实际到访酒藏品尝后购入，"藏葡"坚持只出售值得信赖的"看得见酿酒人的酒"。店内常备约 30 种清酒、200 种葡萄酒，种类颇多。就连供喝酒间隙饮用的以防醉酒的水也非常讲究，目前使用的是鹿儿岛县西酒造用的"宝铭水"。菜单以京都家庭料理为主，另有各种蔬菜及肉类、鱼类菜肴，清酒葡萄酒两相宜。其他一些简单的家常料理也颇受欢迎。店主定期从日本各地邀请清酒生产商举办活动。推荐想要了解清酒相关知识的人参加。

时 17:00—23:30（最后点单时间 23:00）电 03-6264-1759 休 星期日、节假日 址 中央区筑地 1-5-11 AS ONE GINZAEAST 一层 交 地铁筑地站步行 3 分钟 网 www.facebook.com/kurabuutukiji

❶ **KURABUU** (Tsukiji)

Operated by Imadeya, a liquor store in Chiba, KURABUU is a modern *izakaya* (traditional Japanese bar) whose specialty is "liquor that is made in Japan." There are other unique bars and restaurants in KURABUU's neighborhood behind Tsukiji Police Station. The management makes sure that they personally know the producers. Staff are sent to each brewery for tasting. The sight of their regular stock of 30 kinds of sake and 200 kinds of Japanese wines is astounding. The high-quality water, yawaragi-mizu (chaser water), is carefully selected and "Houmei" is currently purchased at KURABUU. The shochu brewery Nishishuzo, in Kagoshima, also uses this particular water regularly as brewing water. The food menu is mainly derived from *obanzai* (Japanese cuisine native to Kyoto) with varieties of vegetables, meat and fish dishes. Simple home-style dishes are also popular. KURABUU holds regular events with sake producers around the country which would suit those who want to deepen their sake knowledge.

(H) 17:00-23:30 (L.C.23:00)　(T) 03-6264-1759　(C) Sun, National holidays　(Ad) 1st floor AS ONE GINZAEAST, 1-5-11 Tsukiji, Chuo-ku　(Ac) 3-min. walk from Tsukiji Station (Tokyo Metro Hibiya Line)
(U) www.facebook.com/kurabuutukiji

照片中部的画作及店铺标志均由曾与卡地亚品牌合作的艺术指导平野杰设计,为店内增添了几抹别样的色彩。葡萄酒展示架的内侧是酒窖。展示架形成的屏障有效减少了空气流通,从而使贮藏状态的葡萄酒可以更好地发酵。

The shop logo and the art work in the center of the photo were created by Tsuyoshi Hirano, an art director who also collaborated with big brands such as Cartier. The undressed concrete wall looks glamourous. Beyond wine shelves on the right is a wine cellar separated by an air curtain.

❷ 麦酒庵（惠比寿）

麦酒庵坐落在可以一览惠比寿地区全景的大厦之中，是位于大塚的一家主打清酒和生蚝的著名料理店的分店，于2014年在惠比寿开业。惠比寿店不卖生蚝，主打美酒，常备约200种清酒及12种精酿啤酒。店主在贮酒上颇为用心，让呈至顾客嘴边的每一口酒都处在最好的状态。店内没有招牌清酒，而是广泛搜罗各种由不同产地出产的口味、香气不同的酒饮以及季节性或限定性酒饮来推荐给顾客。自称清酒侍酒师的店主田中祐晶表示："如果不知道该喝什么好，只要说出你的口味偏好，我们就能为你推荐出最符合你的要求也最适合当日菜肴的那一瓶。"店内料理只使用新鲜食材，菜品以从附近海港当日捕获的鱼产为原材料。"我们希望客人在享用美酒的同时，也能品尝到与之绝配的美食。"

⓽ 星期一至星期五 17:00—23:00，星期六及节假日 15:00—23:00　⓮ 03-3719-3949　⓱ 星期日
⓰ 涩谷区惠比寿南 2-1-5 ES215 大厦 7 层　⓯ JR 或地铁惠比寿站步行 2 分钟　⓲ www.bakushuan.com

❷ BAKUSHUAN (Ebisu)

Located on the top floor of a building overlooking the town of Ebisu, BAKUSHUAN was opened in 2014 as a satellite outlet of a bar in Otsuka which is famous for its sake and oysters. This one, however, is not an oyster bar but has 200 types of sake and twelve types of craft beer to choose from. They carefully keep these drinks in the best conditions for serving. Regarding sake, they select seasonal produce or limited-edition bottles, and wide varieties from various breweries with different flavors and aromas, intentionally avoiding keeping a "regular". "If you do not know what to have, as long as you tell us what you feel like, we can recommend the one that also matches your choice of food," says the master, Yoshiteru Tanaka who calls himself "sommelier" of sake. The kitchen uses only very fresh and recognizable ingredients, mainly seafood from the Ishinomaki and Odawara areas. Yoshiteru continues, "We hope you can enjoy drinking but also the fusion of food and sake."

(H) Mon-Fri 17:00-23:00, Sat & National holidays 15:00-23:00 (T) 03-3719-3949 (C) Sun (Ad) 7th floor ES215 bldg., 2-1-5 Ebisu-minami, Shibuya-ku (Ac) 2-min. walk from Ebisu Station (JR Line & Tokyo Metro Hibiya Line) (U) www.bakushuan.com

店主田中拥有丰富的酒类和料理知识。凡是客人点的酒,他都会提前试饮以便确认状态。"清酒是一种非常敏感的酒,我不想把自己觉得不好喝的酒提供给客人。"

Yoshiteru has extensive knowledge about sake and food. He always tries to taste sake before serving patrons to see if it is in the right condition. He explains, "Sake is a sensitive drink. I hate to serve something which I am not happy with."

麦酒庵

店内提供的料理以各种刺身及煎烤类的和食为主，配合清酒口味特别甄选的芝士拼盘（含3种芝士，1389日元）也很受欢迎。酒单精选中包含清酒和手工啤酒。

Listed on the menu is mostly Japanese cuisine such as sashimi or grilled dishes, typically a good accompaniment for sake. Other items include a cheese platter (¥1,389 for three types), especially selected to go well with sake. The drink menu consists of only sake and craft beer.

❸ twelv.（西麻布）

2016年在西麻布开业的精致清酒吧,理念是"为成熟男女提供一个可以享受清酒的时尚空间"。"twelv."位于半地下室的空间内,一个裸灯泡充当了唯一的外观标识,让人不禁踟蹰:"这里真的是酒吧吗?"半信半疑地推开店门,现代简约的室内设计让人倍感惊艳,这样一家"隐世"小店的内部环境是如此精致典雅。曾操刀普拉达、亚历山大·王等高端品牌门店设计的设计师巧妙运用锡纸等极富未来感的材料,让室内空间看起来极为精致整洁。店内酒单由曾进入品酒大赛前三名的俄罗斯品酒师德米特里·布拉赫拟定,其中最有人气的是一款内含四种清酒的品酒套餐。店内服务内容依时间而变,22点前是全预约制的餐厅,22点后则变身为点单制酒廊,颇有趣。以"Bio"（有机）为关键词的"最新潮"清酒吧"twelv."为客人带来有关清酒及品饮空间的全新体验。

时 20:00— 电 03-6805-0764 休 星期日、节假日 址 港区西麻布4-2-4地下1层 交 地铁六本木站步行12分钟 网 www.twelv.in

❸ twelv. (Nishiazabu)

This is an elaborate sake bar opened in 2016 with the concept of providing a bar where sophisticated people can enjoy sake in style. It is in a semi-basement and the only exterior signage that it is a bar is a naked light bulb. "Is this really a bar?" You may wonder, but behind the door, you will find a cool and modern space that is as minimalistic as possible. Produced by two artists who oversaw Prada and Alexander Wang shops, the impressive décor effectively employs materials such as tin for a simple and lean look. The sake menu was selected by the competent Russian kikisake-shi, Dmitry Bulakh who won the third prize at the 4th World Kikisake-shi Competition. "Kikisake Set" is a popular selection which allows patrons to enjoy four different types of sake. They interestingly change the types of service according to the time; early hours are dedicated to the diners and advance booking is essential, but after 22:00, it is a bar lounge where you can order from their a la carte menu. "Bio" is a key word for twelv.. In this latest sake bar, patrons can experience a new sake with a futuristic atmosphere.

(H) 20:00- (T) 03-6805-0764 (C) Sun, National holidays (Ad) B1 floor, 4-2-4 Nishiazabu, Minato-ku
(Ac) 12-min. walk from Roppongi Station (Tokyo Metro Hibiya Line) (U) www.twelv.in

twelv.

德米特里会根据客人的喜好推荐相应的清酒。对日式料理及日本茶也颇有研究的他日文非常不错,会定期利用开店前的时间举行清酒品鉴活动。

If you do not know an exact type of sake, ask Dmitry and he will select the one that suits your taste. Dmitry's knowledge extends to Japanese food and tea and he speaks excellent Japanese. They organize regular events guided by Dmitry before the bar opens with the aim of enjoying sake.

颇有人气的品酒套餐（4000日元起）。精挑细选的酒器会让清酒的味道更佳。"我们希望客人在品酒的同时也能对日本的美器美物有更多了解，所以店内酒器主要选用日本艺术家或者设计师品牌的作品。"

This is the popular "Kikisake-set" (¥4,000-). They use selected sake vessels to bring out the best flavor of each sake. "My aim is to introduce customers to sake and also other things of Japanese beauty. The sake vessels we use are mostly made by Japanese artists or designer brands."

酒单内全部是德米特里精选的有机清酒及以清酒为基酒调制的鸡尾酒（另有非酒精饮料）。清酒酒单会按季节更替。

Alcohol beverages on the menu are only organic sake selected by Dmitry and cocktails made with these sake (non-alcoholic drinks are also available). They seasonally change the selection of sake.

店内料理尽可能只用日本本土有机食材。包括时令蔬菜、咸菜、酱腌豆腐、芝士等在内的全部菜品，都非常适合搭配清酒进食。

They also use as much organic and domestic produce as possible to prepare food. All the items on the menu would be perfect to have with sake, including plates of seasonal fresh and pickled vegetables, miso-pickled tofu, and cheese.

与千叶县木户泉酒造联名推出的"twelv."原创气泡清酒。为配合"twelv."品牌形象,酒瓶外部设计也十分简洁,颇具现代感。限定 1 万瓶,GINZA SIX 及伊势丹有售。

Limited to 10,000 bottles, twelv.'s original sparkling sake was co-produced with Kidoizumi Brewery in Chiba. The packaging was stylishly designed to suit the image of the bar. It is sold at GINZA SIX and Isetan.

除吧台座席外另有餐桌座席。店铺位于毗邻马路的"The Wall"大厦半地下室内,若初次前往,建议在预约时和店员确认好具体位置。

There are tables as well as a counter. It would be very difficult for first-time visitors to find twelv. because it is in the semi-basement in a building facing a busy road. Ask for their exact location when booking tables.

店内设有 DJ 台。在俱乐部般的氛围中享用清酒是此店最大的魅力之一。客人在这里可以收获不同于其他清酒酒馆的全新饮酒体验。

Another attraction of twelv. is a DJ booth. They offer a whole new experience of appreciating sake within a club-like atmosphere, unlike other Japanese sake bars which have a traditional setting.

❹ 希纺庵（池袋）

坚持只提供清酒的酒吧。位于大厦6层，大落地窗让店内空间开放感十足。店主渡边元康希望将清酒、客人与食材紧密联系在一起，"织成网"，故为店取名"希纺庵"。渡边在22岁初次品尝清酒美味，因痴迷于清酒，便选择在百货商场的清酒卖场就职，9年前于池袋创业开店，6年前搬至现址。店内常备8种严选清酒，每两周更换。"我们希望客人能以开放的心态尝试各种清酒，所以在正常的1合（180毫升）外也接受半合及60毫升的点单。"店内提供的酒饮全部来自可靠的酿酒师，想要发自内心地推荐给客人尝尝是最重要的进货标准。店主会利用休息日亲自走访日本各地酒藏，到了酿酒时节，还会去相熟的酒藏给别人帮忙。菜单里为搭配清酒而特制的各类下酒菜无一不美味。每来这里一次对清酒的认识都会更深一层。

时 星期二至星期六 17:00—23:00（最后点单时间 22:30），星期日 15:00—21:00（最后点单时间 20:30）
电 03-3987-7518 休 星期一 址 丰岛区西池袋 3-31-15 皇家广场 II 6 层 交 JR 池袋站步行 5 分钟
网 无

❹ KIBOUAN (Ikebukuro)

So passionate about sake, the owner made this a "sake-only" bar. It is on the 6th floor, with a big window giving it a spacious feel. Ki-bo-an (literally a house to weave hopes) was named by the owner, Tomoyasu Watanabe, to provide a space to interweave sake, people and food, etc. Having realized the deliciousness of sake at the age of 22 and gained experience at a liquor section in a department store, he opened a bar 9 years ago here in Ikebukuro. This is his 6th year at the current location. The bar regularly carries 8 different kinds of sake, half of which are changed to different kinds every 2 weeks. "Since I'd like customers to taste many types, we serve sake in a half-*go* (standard measurement of sake; half-go is 90 ml) or even 60 ml," says Tomoyasu. He makes sure to buy sake from trusted breweries and choose what he wants customers to try. On his days off, he visits breweries around the country and sometimes helps them in their production stage. The food here is also appetizing. You would taste and learn something new about sake with every visit.

(H) Tue-Sat 17:00-23:00 (L.C.22:30), Sun 15:00-21:00 (L.C.20:30) (T) 03-3987-7518 (C) Mon (Ad) 6th floor, Royal Plaza II, 3-31-15 Nishi-Ikebukuro, Toshima-ku (Ac) 5-min. walk from Ikebukuro Station (JR Yamanote Line) (U) N/A

店内的背景音乐总是爵士乐。以吧台为中心，店内布置精致优雅。下酒菜中，以黄油烤扇贝为代表的海鲜类料理最有人气。鱼肉饼、香肠、芝士类的加工食品全部来自值得信赖的制造商。

Jazz is played in the bar. Stylish décor and the counter at the center give this bar a refined atmosphere. Butter-grilled scallops and other seafood items are popular dishes. Processed food such as fish cakes, sausages, and cheese are purchased from reliable producers.

❺ SOREGASHI（五反田）

"SOREGASHI"在五反田区域拥有"酒场 SOREGASHI""鸡肉料理 SOREGASHI"等多家分店,每家都有各自的特色。其中,擅长清酒和牛肉搭配的是"肉料理 SOREGASHI"料理店。乍看之下,肉和清酒似乎"毫无干系",但这家店可以刷新你我的认知。所有清酒以最适合各款肉类料理的温度呈现,让人觉得这样的搭配恰到好处。店内每季供应 60 种不同清酒,以纯米酒为主,店主可根据客人喜好提供推荐。"搭配肉类料理饮用的清酒最好口感丰富或酸味较强,比起口感清淡的酒类,口感丰富的清酒与料理在口腔内会产生更美妙的碰撞。"自称"进酒师"的店长山本贵文如是说。想品尝 A5 级黑毛和牛的话,除了烤牛肉、寿喜锅等单品料理,也推荐点套餐。套餐从前菜到主菜、锅物料理,再到最后的全食,全部以肉类为主,让人吃后大感满足。"肉类可以进一步激发清酒的美味,产生不同于单独饮酒时的全新口感,请一定要试试。"

⏰星期一至星期六 17:00—次日 1:00,星期日、节假日 17:00—24:00 ☎03-6420-3092 (休)无 (址)品川区西五反田 1-4-8-202 (交)JR 五反田站步行 3 分钟 (网)www.soregashi.jp/niku

❺ **SOREGASHI** (Gotanda)

Soregashi operates several bars in the Gotanda area and each one, for example Sakaba Soregashi (sake) and Tori-Ryori Soregashi (chicken), has its own uniqueness. This bar, Niku-Ryori Soregashi offers the fusion of beef and sake to the fullest. Do not worry about the mix of sake and beef. Sake is served at the perfect temperature for each beef dish. From their collection of 60 kinds of seasonal sake with *Junmai-shu* (sake made from only rice, water, and koji) at the center, they can serve you with the one that suits your tastes if you ask. "To go with beef, choose the one that is rich in *umami* (the fifth taste sensation) and acidity rather than lighter ones to enjoy the pairing of flavors," says the manager, Takafumi Yamamoto who calls himself a "sake leader." Here, the highest quality wagyu beef (A5 grade) is used for the roast, shabu-suki hotpot, and course meals with arrays of beef-centered dishes from the starter, to the main course, to the rice at the end. "Beef enhances the umami in sake, then the flavor of sake changes dramatically. Please come and let us introduce that to you."

(H) Mon-Sat 17:00-01:00, Sun & National holidays 17:00-24:00　(T) 03-6420-3092　(C) None　(Ad) 1-4-8-202 Nishi-Gotanda, Shinagawa-ku　(Ac) 3-min. walk from Gotanda Station (JR Yamanote Line)
(U) www.soregashi.jp/niku

SOREGASHI

据说，当肉与清酒的温度接近时可以激发彼此最美妙的口感。店内除提供清酒外还提供红酒，设有吧台座席、沙发座席及包房。餐厅位于目黑川沿岸，红绳暖帘的广告牌是唯一标识。

Serving beef dishes and sake at a similar temperature is the key to maximize the synergy of taste, says the manager. They regularly change their stock of sake every season. Wines are also available. It is on a street along the Meguro River, with a small signboard of their logo depicting *Nawa-noren* (a short curtain made with ropes hung at a shop entrance). Inside, they have a counter, booths and private rooms.

SOREGASHI

❻ 银座君嶋屋（银座）

这家店是拥有 120 年历史的横滨老牌酒类专卖店横滨君嶋屋在银座的分店，于 2013 年开业，坐落在银座及东京站之间的繁华区域中，周边有许多日本各都道府县特产直销商店。银座君嶋屋主营各种适合配餐饮用的清酒，种类繁多，百饮不厌。除清酒外，店内还出售烧酒、红酒及精酿烈酒，银座一带的许多餐厅都是在这里挑选酒品。店内还设有立席吧台，提供品酒套餐及 10 种以上清酒。酒单中最受欢迎的当属店主君嶋哲至亲自酿造的原创酒品"情热系列"。佐酒小菜自然也不含糊，从全国各地精选的芝士拼盘、酱渍芝士、生火腿等，无论原材料还是味道都是百里挑一。来自大阪著名章鱼烧店"Takoriki"的章鱼烧尤其适合搭配清酒或红酒享用。店内光线明亮，氛围轻松，适合爱酒之士进去小酌一杯。

时 星期一至星期五 10:30—21:00，星期六 10:30—20:00，星期日、节假日 10:30—19:00 电 03-5159-6880 休 无（除年末年初外） 址 中央区银座 1-2-1 绀屋大厦 1 层 交 JR 有乐町站步行 4 分钟 网 www.kimijimaya.co.jp

❻ GINZA KIMIJIMAYA (Ginza)

In 2013, Yokohama Kimijimaya, a liquor wholesaler with more than 120 years of history, opened this shop in the area between Ginza and Tokyo Station, where many shops promoting local specialties around the country are situated. Many of Kimijimaya's sake go well with various foods so you will never become bored. Many restaurants in Ginza also shop here because of the large stock of liquors such as sake, shochu, wines and craft spirits. Tasting is available at the bar too. Choices are a kikisake-set or a la carte of 10 or more sake. Popular ones are "*Jonetsu* (passion)" , series sake created by the president, Satoshi Kimijima, who personally visited and commissioned brewers to produce for Kimijimaya. The snacks are also alluring, quality and tasty food from all over Japan. The menu includes a cheese platter, miso-pickled cheese, prosciutto, and above all, *tako-yaki* (octopus balls) from Takoriki, Osaka, which goes well with both sake and wine. The shop has a bright atmosphere that welcomes all for a drink.

(H) Mon-Fri 10:30-21:00 Sat 10:30-20:00 Sun & National holidays 10:30-19:00 (T) 03-5159-6880
(C) None (except for New Year holidays) (Ad) 1st floor, Konya Bldg., 1-2-1Ginza, Chuo-ku
(Ac) 4-min. walk from Yurakucho Station (JR Yamanote Line) (U) www.kimijimaya.co.jp

❼ 采（三轩茶屋）

由三轩茶屋的著名烧鸟（烤鸡肉串）店——"床岛"运营的站席清酒专门店，坐落在紧挨车站的小巷里。人气极旺，每天一开门就坐满了食客。店内供应六七十种清酒，具体喝哪种既可以请店员推荐，也可以按黑板上列出的四种口味（熏酒、爽酒、醇酒、熟酒）自行挑选。"建议不知道该喝什么的人，先把四种都试试，比较一下，就会明白不同土地产出的不同大米酿出的酒之间的不同。"店长三轮英司介绍道。低至半合的最小点单量实在让人开心，每种少喝一点，喝的种类就可以多一点。一般餐厅在上酒前先上的小菜在这里是鸡汤粥，用来酒前开胃也好，一口一口地佐酒喝也适宜。为独自来店的客人特别提供的六种一盘的单人佐酒套餐（含税 800 日元）颇受欢迎。

⓪ 18:00—次日 2:00（最后点单时间 1:30）⑨ 03-6453-4511 ⑭ 星期日 ⑱ 世田谷区三轩茶屋 2-13-9 ⓧ 东急田园都市线三轩茶屋站步行 5 分钟 ⑳ 无

❼ SAI (Sangenjaya)

Sai is a standing-only sake bar, operated by Tokoshima, a famous *yakitori* (grilled chicken skewers) restaurant in Sangenjaya. Located in a small alley close to the station, it is so popular that it fills up as soon as it opens each evening. There are always 60 to 70 kinds of sake to choose from. Staff can help you decide what to have or you can choose from suggestions on the board, written in 4 groups: *Kunshu* (fragrant and clean), *Soshu* (clean and moderate aroma), *Junshu* (rich and moderate aroma), *Jukushu* (aged rich and fragrant). "If you are still unsure, try all four types and find out how they differ from each other depending on the rice and soil," says Eiji Miwa, the bar manager. It helps that you can order sake from as little as 90 ml so that you can taste many kinds. It is their specialty to serve rice porridge in chicken broth as *otoshi* (a mandatory appetizer). How you eat it is your choice; finish first or eat slowly with sake. A popular dish for a single customer is Ohitorisama-set (¥800, incl. tax) with 6 kinds of amuse-bouche on a plate.

(H) 18:00-2:00 (L.C.1:30) (T) 03-6453-4511 (C) Sun (Ad) 2-13-9 Sangenjaya, Setagaya-ku (Ac) 5-min. walk from Sangenjaya Station (Tokyu Den-en Toshi Line) (U) N/A

❽ TSUNEMATSU 久藏商店（月岛）

月岛地区以特色料理"文字烧"而闻名于世。常松治郎在 2016 年为推广岛根县特产酒而开的居酒屋就坐落在月岛车站附近。进入店内，首先映入眼帘的是一张巨大的天然榉木餐桌，架子上装饰的各种一升装酒瓶将大大的餐桌衬托得格外引人注目。店内供应 50 余种来自日本各地的清酒，以岛根县清酒为主。酒单外还有一些当季精选的"隐藏酒单"，可以告诉店家自己的喜好让他推荐。海鲜料理、岛根乡土料理，以及"成年人的鸡蛋豆腐"（999 日元）、"芜菁薄切生肉"（555 日元）这类乍看菜名有些奇怪的下酒菜，可以为客人带来味觉、视觉双重满足。店内氛围轻松，既可独自一人小酌，也适合与公司同事或志趣相投的伙伴随意喝上几杯，有时还会有女孩子们相约来此聚会，算是一个在闲暇之余品味清酒的好去处。

⏰星期一至星期五、节假日前日 17:00—00:00（最后点单时间 23:30），星期六、星期日 15:00—23:00（最后点单时间 22:30）☎03-6204-9740 休 不固定 址 中央区月岛 1-6-12 交 地铁月岛站步行 1 分钟 网 无

❽ TSUNEMATSU KYUZO SHOTEN
(Tsukishima)

Tsunematsu is located one minute away from the railway station of Tsukishima, the *monja-yaki* (savory pancakes) town. The bar was opened by Jiro Tsunematsu in 2016, with the aspiration of introducing the local sake from Shimane to Tokyoites. Once in the bar, you cannot help noticing the huge table made of solid Japanese zelkova timber. Around the table and on the shelves, large sake bottles are displayed to set off the table. They regularly keep about 50 kinds of locally brewed sake with the ones from Shimane at the center. Other than that, they have specials from time to time, so it would be a good idea to tell them your desired taste and let them choose for you. The food includes fresh seafood and Shimane's local fare. There are also original snacks such as Otona-no-tamago-tofu (savory egg custard with lavish toppings) (¥999) and turnip carpaccio (¥550) to feast one's eyes as well as taste buds. All kinds of customers come to this bar. Whether you are on your own or in a group of colleagues/friends, it is a rare spot to enjoy sake in a relaxed style.

(H) Mon-Fri, and the day before National holidays 17:00-0:00 (L.C.23:30), Sat-Sun 15:00-23:00 (L.C.22:30) (T) 03-6204-9740 (C) Irregular (Ad) 1-6-12 Tsukishima, Chuo-ku (Ac) 1-min. walk from Tsukishima Station (Tokyo Metro Line, Toei Subway) (U) N/A

❾ 未来日本酒店 DAIKANYAMA（代官山）

日本全国有 1500 多家酒藏，然而在东京流通的清酒几乎全部来自几家有限的大型酒藏。为了让更多人认识日本那些地方性的工艺精湛充满魅力的小型酒藏，品尝到年青一代酿酒师酿造的"新感觉"清酒，"未来日本酒店 DAIKANYAMA"于 2017 年在代官山诞生。这家店在开业之初只做网店，渐渐地通过在商场举办活动及开展快闪店扩大了品牌的知名度。店内约 180 种酒饮标价每杯 400 日元起，店主希望客人可以像听音乐或试衣服一样，以更低的门槛尝试并享受多种多样的清酒。除清酒外，店内还提供众多的甜酒、烧酒以及以梅酒为代表的利口酒，也全部产自小型酒藏。店主会不定期邀请各地的酒藏主或酿酒师来到店面举行各种活动，对清酒有兴趣的人一定要来这里试试。

时 13:00—22:00　电 03-6312-2448　休 无　址 涩谷区代官山町 14-11　交 东急东横线代官山站步行 3 分钟　网 miraisake.com

❾ MIRAINIHONSHUTEN DAIKANYAMA
(Daikanyama)

There are more than 1500 breweries in Japan, but most of the sake marketed in Tokyo is produced by large-scale brewers. Mirai-nihonshu-ten's concept was to promote more new generation sake from smaller but interesting local breweries with good craftsmanship or by younger brewers. The bar was opened in 2017: Originally it had started as an online shop but since then has gradually expanded its business through special events in shopping malls or pop-up retailing. Tasting is available from ¥400 per glass from a selection of 180 kinds of sake. The management wants this bar to be a place where customers can readily try sake in the same manner that they sample music or try on clothes. Other types of beverages available here include shochu, and liqueur such as *ume-shu* (plum liqueur) and *amazake* (sweet fermented rice drink) produced by sake brewers. Events with brewers or distillers are held from time to time. Those who have never tasted sake, or who are interested but have no idea where to start, are especially welcome to visit this bar.

(H) 13:00-22:00　(T) 03-6312-2448　(C) None　(Ad) 14-11 Daikanyama-cho, Shibuya-ku　(Ac) 3-min. walk from Daikanyama Station (Tokyu Toyoko Line)　(U) miraisake.com

清酒小知识 1
日本文化与清酒

历史

著于公元 3 世纪的中国史书《三国志·魏书·倭人传》曾云："日本人,即倭人,嗜酒。"奈良时代(公元 710—794)编撰的日本史书《日本书纪》中,记载了神话人物须佐之男在饮用了果酒或以近似于清酒制法酿造的某种酒之后,用剑降服了传说中的巨型蛇怪八岐大蛇的故事。这段神话传说相关的更为具体的记载出现在与《日本书纪》同时期的《播磨国风土记》中,里面清晰记述了清酒的酿造方法与原料,《播磨国风土记》中的这段记载目前被认为是有关清酒最古老的文字记录。后来,酿酒技术在日本各地被推广开来,日本人历史悠久的饮酒风俗也传承至今,全日本现在有近 1500 家酒藏。

仪式

在婚丧嫁娶、庆典祭祀上,日本各地都有其独特的风俗习惯。不过无论有怎样的习俗,这其中一定都少不了"清酒"。这是因为这些仪式很多需要在神社内进行,而清酒是供品中必不可少的。某些地方甚至还有专门供奉酿酒神的神社。

四季

在一年之中,日本人有很多"不得不"饮酒的时刻。大家都知道春天赏樱要喝"花见酒",秋季望月该饮"月见酒",冬日观雪也不能忘了"雪见酒",其实每年 6 月的最后一天(即晦日)还要喝"越夏酒",意寓洗去半年来积攒的一身尘埃。从古至今,酒与风俗仪式、四季变换始终关系密切。

TRIVIA ON SAKE_1
JAPANESE CULTURE AND SAKE

HISTORY

In Gishi-wajinden, a part of a Chinese history book in the 3rd century, there was a description that wajin (Japanese) "enjoy drinking sake." In the *Nihon Shoki* (Chronicles of Japan), which was compiled in the 8th century, it is written that *Susano-o-no-mikoto* (a Japanese mythological god) cut down *Yamatano-orochi* (a mythological monster) which drunk either fruit liquor or alcohol brewed with a similar method to sake-making. Also in *Harima Fudoki* (local gazetteer from a region in western Japan), which was written at around the same period as *Nihon Shoki*, it said that sake was produced from rice. This is presently regarded as the oldest written account on sake. Since then, sake-production has spread all over Japan and the current number of sake breweries is just under 1500. There is no doubt that sake drinking culture begun in ancient times and that tradition has been inherited through history.

EVENT

Weddings, funerals, local festivals and other traditional events are held in a dignified manner which can be different from region to region, but sake is a must on any occasion. It is thought that sake is an essential offering in these rituals which are done as a dedication to the Shinto gods. Some Shinto shrines worship the god of sake-brewing.

FOUR SEASONS

In Japan, there are various occasions in which sake is customarily offered. *Hanami-zake* (sake to drink at cherry blossom-viewing parties), *Tsukimi-zake* (sake to drink while viewing the moon), and *Yukimi-zake* (sake to drink while viewing snow) to name a few. Another example is *Nagoshi-no-sake* which is drunk on the last day of the 6th lunar month as part of the half-yearly purification ritual. Sake has long been inseparable from not only annual events but also seasonal celebrations.

第一章

第二章

品日本酒：
那些魅力主厨经营的小店

DRINK:
RESTAURANTS WITH
WELCOMING OWNER-CHEFS

在与主厨交流的同时小酌一杯,
正是清酒的魅力之处

可能因为父亲曾经营日本料理店,Yoshiro 从小就对食物感兴趣,只不过,那时候没想过要把料理当成工作。在立志成为料理人后,他不仅去上了餐饮协调员培训学校,参加清酒讲座,一点点从料理的基础知识学起,还考取了许多品酒师之类的餐饮相关资格证书。Yoshiro 平时最爱去的就是那些充满魅力的主厨所开的店,如果一家店"不光酒美味,料理也好吃"的话他就会常去。"米和水的品质基本上决定了清酒的口味,所以对于出售酒品的餐厅而言,料理与酒品合适的搭配就显得非常重要。跟主厨商量怎么选酒和料理,绝对错不了。"找到一个可以与主厨建立信赖关系的料理店,会让你在享用美酒美食时更加充实。

第二章

Tasting sake over a chat with the owner-chef is one of the enjoyments of sake

Yoshiro's interest in food from a young age can be attributed to his father who had run a Japanese restaurant. "But I didn't think that I would end up working in the food industry," says Yoshiro. After deciding to become a cooking specialist, he spent his days attending food coordinator schools and sake seminars to study from the basics. He has obtained multiple food certifications including kikisake-shi. He likes to frequent a particular restaurant with a welcoming owner-chef. He explains, "good food, not just good sake," is another contributing factor for him returning to a restaurant. "The flavor of sake is largely dependent on the quality of the flavor of the rice and water used to make it, which is why its chemistry with the accompanying food is incredibly important. Discussing sake and food with the owner-chef is the best way to choose the right combination." Finding a restaurant with a reliable owner-chef whose palate matches your own is sure to guarantee a more enjoyable time with sake and food.

ARTRIP ADVISER
艺术之旅顾问

Yoshiro

料理家、铁人三项全能运动员。受料理人父亲的影响走上料理之路。拥有清酒学讲师、品酒师、蔬菜师等资格证书。最近活跃在电视及杂志领域。负责线上及线下料理教室的运营。

A cooking specialist and triathlete. He started his journey as a cooking specialist due to his father's influences. Certified as a vegetable sommelier, and as both a lecturer and Kikisake-shi by the Sake Service Institute. He has appeared in various TV programs, and also runs cooking seminars and online classes.

❿ 青（代代木上原）

在代代木上原站附近有一间小店，门前垂着一大张暖帘，青底白字，单书一个"青"字。店主山根崇义曾长年在料理研究家 Yoshiro 的父亲身边学习。他在结束学徒生涯后自立门户，"青"于 2002 年开业，多年来广受赞誉，食客络绎不绝，在代代木上原一带的料理店中保持着稳固的地位。山根为店取名"青"字是为了时刻提醒"自己不要忘记曾经如蓝色天空般纯洁的初心"。店内供应的 30 多种酒饮基本都有着山根偏爱的"爽快又厚重"的口感，这样的酒适合作为餐中酒配餐饮用。料理主要以鱼类料理为主，使用的都是当日采购的新鲜食材。菜单中可以看到许多经典日式料理的"改良版"，单品菜单每月更新。"我就在吧台里为食客全力以赴地做菜。关于酒或料理有什么问题的话，直接问我就好。"

时 星期二至星期六 18:00—次日 2:00，星期日 18:00—0:00　电 03-3485-8808　休 星期一　址 涩谷区西原 3-24-12　交 小田急线代代木上原站步行 5 分钟　网 无

❿ **AO** (Yoyogi Uehara)

This restaurant is a few minutes' walk from Yoyogi Uehara station and is recognized by the sign made of large blue cloth hanging down from the restaurant with the logo *"Ao"* (meaning "blue" in Japanese) illustrated in white. The owner opened it in 2002 after working as an apprentice at cooking specialist Yoshiro Takahashi's father's restaurant. Ao has since maintained its position as one of the most popular restaurants in Yoyogi Uehara. Owner Takayoshi Yamane named it Ao to remember that he is "still learning and needs to keep the heart of a novice, like a blue cloudless sky." Takayoshi likes sake that is "crisp but with depth," and always keeps about 30 varieties based mostly on his taste. "The flavors of the sake should complement the food." The menu is composed mainly of original dishes cooked using fish purchased at the *Tsukiji fish market* (a huge fish market in central Tokyo) every morning. Most dishes are standard Japanese dishes rearranged in the owner's style, and the regular menu changes at least once a month. "I prepare my dishes behind the counter with great care. Please feel free to ask me questions about the sake and the dishes when you visit."

(H) Tue-Sun 18:00-2:00 (closes at midnight on Sundays) (T) 03-3485–8808 (C) Mon (Ad) 3-24-12 Nishihara, Shibuya-ku (Ac) 5-min. walk from Yoyogi Uehara Station (Odakyu Line) (U) N/A

一进门就是吧台。共有 8 个座位，所有座位都可以清楚地看到吧台内部。常有熟客会专为见店主山根一面而来。店内的半包包间内还有两张餐桌座席，常被用来举办同好会或公司聚餐。

There are 8 seats at the counter by the entrance. The inside of the counter can be viewed from every seat. One of the attractions for most regular customers is to visit Takayoshi. There is also a semi-private room in the back with 2 tables, which is often used for small groups and company gatherings.

餐桌座席被笼罩在一片柔和灯光之下,为享用美酒佳肴提供了一个绝佳环境。单品及迷你套餐很受欢迎。迷你套餐(3400日元)含前菜、炸物等7道料理,"除此之外,还可以尝尝饭类料理"。

The table seats are lit with soft lights providing an atmosphere that lets you enjoy sake and food. Both the a la carte menu and the 7-course meal (¥ 3,400), which includes an amuse-bouche and a deep-fried dish, are very popular. Takayoshi adds, "If that is not enough, please try one of our rice dishes."

店内提供的酒品多为冷酒,若客人有要求也可加热。当然,店内也出售本就适合加热的酒饮。酒单基本不会出现变动,偶尔会出现一些当季推荐的特别酒单。

Most of the sake is the type that should be served at room temperature, but can be warmed upon request. There is also sake that is intended to be served hot. The basic lineup stays the same, but seasonal recommendations are also often available.

第二章

店内用的酒器大部分出自日本名家之手,其中部分对外出售。"多少也有些支持这些艺术家的想法,看到觉得不错的酒器就买回来用在店里。"所有酒饮以 1 合为最小单位出售。

Most of the china used in the restaurant is made by Japanese artists, and some are sold in the restaurant "to help promote the artists. If I like a certain type of china, I use it in my restaurant frequently." The minimum serving size for sake is 1-go (180ml).

鲭鱼和香菇的搭配再加上醋味渍物的"锦上添花",这些食材碰撞出了店内的招牌料理 —— 醋腌烤鲭鱼香菇(800 日元)。使用生活中常见的食材,将其重新搭配组合出让人出乎意料的美味正是"青"料理店一直在追求的料理理念。

A popular dish is the pickled grilled mackerel and shiitake mushrooms (¥ 800), which is a surprising combination to be mixed with vinegar. Combining classic ingredients with unusual seasoning to create a unique dish is Ao's specialty.

AO

⓫ 松（下北泽）

下北泽地区汇聚了许多小剧场、古着店及杂货店，颇受年轻人喜爱。和食居酒屋"松"位于下北泽中心区域之外，店铺显露出一股别样的成熟风韵。2004年开业的"松"，菜式以传统和食为主。料理品种丰富，菜单每月更替。店内供应的二三十种清酒大多都是纯米酒，因为店主松崎英智认为："纯米酒口感更厚重，更有喝酒的感觉，与和食的契合度也高。当然如果客人喜欢口感更轻盈的酒，我也可以推荐。"对每一道呈现在食客餐桌上的料理，松崎都毫不懈怠，"客人品尝时的惊艳表情让一切努力都变得值得"。近来人气最高的料理是热土豆沙拉和炖煮豚骨。店内配有吧台座席、普通餐桌座席以及可作为半包包间使用的餐桌座席，可以满足食客的多种需求。店主自豪地说："欢迎所有喜爱美酒美食的朋友光临'松'。店内不光有清酒，还有种类丰富的烧酒。"

时 18:00—次日 2:00　电 03-3418-2778　休 星期四　址 世田谷区北泽 2-15-11　交 小田急线或京王井之头线下北泽站步行 4 分钟　网 02shimokita.favy.jp

⓫ MATSU（Shimokitazawa）

Shimokitazawa is an area of many small theatres and shops for young people, such as second-hand clothing shops and general stores. A little distance away from the hustle and bustle is the stylish Japanese izakaya "Matsu." It opened in 2004 serving mainly traditional Japanese food with an extensive menu of everything from fish to meat and vegetables, as well as a monthly menu. There are always about 20 to 30 types of sake available. Most of the sake is Junmai-shu, as owner-chef Hidetomo Matsuzaki believes, "Junmai-shu is solid. It also goes well with Japanese cuisine. But if you prefer lighter sake, please let me know." Hidetomo's motto for all of his dishes is to always add a twist, so that "the customers look surprised and enjoy the dishes even more. That is what makes it worthwhile." Recent popular dishes are the warm potato salad and braised *tonkotsu* (pork short ribs). There are counter seats, table seats, and a semi-private room. "I welcome everyone who loves drinking and eating. We also have many types of shochu."

(H) 18:00-2:00 (T) 03-3418-2778 (C) Thu (Ad) 2-15-11 Kitazawa, Setagaya-ku (Ac) 4-min. walk from Shimokitazawa Station (Odakyu and Keio Inokashira Line) (U) 02shimokita.favy.jp ※Please note that services are not available in English.

⑫ SEKI 亭（二子玉川）

SEKI 亭位于二子玉川站附近。店内面积不大，只有一张用名贵京都木材制作的 10 人座吧台及一间全包包房，狭长的店面布局是为了让店主关齐宽可以随时照顾到店内的所有客人。在这里，客人可以配合着传统手握寿司、天妇罗以及用浓厚的日式高汤做成的各类日式料理，享用日本各地的特色地方酒。招牌天妇罗，制作虽简单却饱含了店主对料理的满腔热忱。料理的原材料全部是当日一早从筑地市场选购，只为让客人品尝到最新鲜的味道。店内日常供应 30 种以上精选清酒，除搭配和食饮用的餐酒外，还有各种季节限定的以及部分极难入手的"极品"清酒。在酒饮选择上，店主不会做硬性推荐，根据你自己的口味和偏好选择即可，酒单每周一换。店主在料理、店内装潢、待客、饮酒器具的选择上都颇下功夫，让人切身体会到日式美学的奥妙。适合私人聚会、商务宴请等各种场合。

时 17:00—23:00（最后点单时间 22:30） 电 03-6432-7867 休 不固定 址 世田谷区玉川 3-9-1 第三明友大厦 2 层 交 东急田园都市线二子玉川站步行 5 分钟 网 www.natomics2010.com

⓬ SEKITEI (Futakotamagawa)

"Sekitei," located at a 5-minute walk from Futakotamagawa station, is a small restaurant with 10 counter seats made of precious wood from Kyoto, and has 1 private room. The restaurant serves traditional *nigirizushi* (hand-made sushi), single orders of *tempura* (deep-fried battered seafood and vegetables), boasts various dishes with select *dashi* (broth), and each dish can be enjoyed with local sake from various parts of Japan. Its scale enables owner-chef Hiro Sekinari to see all corners of the restaurant as he meticulously prepares the freshly deep-fried tempura and other dishes. The ingredients are all seasonal with the freshest ingredients from all over the country purchased at Tsukiji fish market every morning. The sake, which is chosen from all varieties including seasonal sake and rare sake, brings out the flavor of the food, and changes weekly while maintaining a variety of 30 or more. The food, atmosphere, customer service and china are designed to show the goodness of Japan. It is suitable for private dinners, client dinners, and other purposes.

(H) 17:00-23:00 (L.C.22:30)　(T) 03-6432-7867　(C) Irregular　(Ad) 2nd floor No.3 Meiyu Bldg. 3-9-1 Tamagawa, Setagaya-ku　(Ac) 5-min. walk from Futakotamagawa Station (Tokyu Denentoshi Line)　(U) www.natomics2010.com

SEKITEI

⑬ 旬菜 SUGAYA（千岁船桥）

颇受附近居民喜爱的居酒屋"旬菜 SUGAYA"坐落在商店街与居民区中间的区域，从小田急线千岁船桥站步行4分钟即可到达。刺身拼盘是店内招牌料理，料理中所使用的食材都是店主每日一早从市场购回的，保证鲜味十足，一盘400日元左右，其分量之大、价格之实惠让人简直不敢相信，难怪许多熟客常来光顾。酒单上的清酒一般维持在20种左右，一种只有一瓶，喝完了就再开一种新酒。"我们主要从几家值得信赖的酒藏挑选适合当季料理的清酒，尤其是适合当季海鲜的清酒。饭和清酒的原料都是大米，所以适合配米饭的料理同样也适合搭配清酒。选酒的第一要义是酒与料理的配合度，而非它的名贵性或稀少性。"店主菅谷亮介绍道。在这里，客人可以以最实惠的价格品尝到店主认真挑选的清酒以及精心制作的鱼类料理。

(时) 17:00—24:00（最后点单时间 22:30）(电) 03-6413-5450 (休) 星期三 (址) 世田谷区船桥 1-7-3 信长君之台所大厦 3 层 (交) 小田急线千岁船桥站步行 4 分钟 (网) 无

⓭ Shunsai SUGAYA (Chitose Funabashi)

"Shunsai SUGAYA" is a well-established local izakaya at a 4-minute walk from Chitose Funabashi station on the Odakyu line, and is an area midway between the shopping district and the residential district. The izakaya boasts cuisine made from fresh seafood purchased from the market every morning. All of the *sashimi* (sliced raw fish) dishes are amazingly generous for the price at around 400 yen per dish. This explains why so many customers come here. There are about 20 types of sake always available, and a new brand is opened once a bottle is finished. "I mainly choose sake made by breweries that I trust. Dishes such as *otsumami* (small dishes and snacks eaten with alcoholic beverage) that go well with rice also match well with sake. The fundamental thing is to choose sake that goes well with the food rather than selecting flamboyant ones," says owner-chef Ryo Sugaya. The greatest pleasure at this restaurant is being able to enjoy specially selected sake with fish prepared meticulously, all at a reasonable price.

(H) 17:00-24:00 (L.C.22:30) (T) 03-6413-5450 (C) Wed (Ad) 3rd floor Nobunagakun-no-daidokoro Bldg. 1-7-3 Funabashi, Setagaya-ku (Ac) 4-min. walk from Chitosefunabashi Station (Odakyu Line) (U) N/A

⑭ 四季饭晴间（代代木八幡）

这是一家隐藏在安静住宅区的小众餐厅。店主中原知美持有侍酒师及品酒师资格证，因热爱料理而辞去了原本的职员工作，自己开了一间餐厅。中原喜欢对传统日式料理加以改造，创作出和洋结合的全新料理。基本菜单内包含四组套餐，享用了套餐后，客人可根据自身饱腹程度再加点鲷鱼土锅饭或单品料理。料理全部使用当季食材制作，不添加化学调料、不过度调味的烹调方法让食材的味道得以充分发挥。荤素食材均衡搭配，店主中原小姐在研发料理时也考虑到了客人的身体健康。古董菜碟、筷子架、酒器就放在餐桌抽屉里，客人可按个人喜好选择。店内提供的清酒全部产自店主的老家福井县，有的清酒与和食配合度高，碰撞出的味道令人惊艳，有些清酒是季节限定，甚至还有一部分来自小型私人作坊，东京地区鲜有流通。除清酒外店内还提供手工酿造的葡萄酒、家酿果酒等市面上难以入手的酒饮。店内氛围舒缓，让人不禁想要在这里与美酒共度愉悦一刻，放松身心。

⑭ 星期二至星期六 18:00—23:00（最后点单时间 22:30），星期日 17:00—22:00（最后点单时间 21:30）
⑭ 03-6804-9980 ⑭ 星期一 ⑭ 涩谷区上原 1-1-20 ⑭ 地铁代代木八幡站步行 1 分钟 ⑭ ha-re-ma.com

⑭ SHIKIGOHAN HAREMA (Yoyogi Hachiman)

This restaurant is in a quiet residential area, almost hidden away. The owner-chef, Tomomi Nakahara, a former company worker who developed a penchant for cooking, is a certified sommelier and kikisake-shi. She uses Japanese cuisine as a base to create Japanese-western fusion cuisine. The menu includes 4-course options. The sea bream and rice cooked in a clay pot or other dishes can also be additionally ordered. All dishes incorporate seasonal ingredients in their natural flavors and without excessive seasoning. They are healthy with plenty of vegetables. The tables have drawers containing antique plates, chopstick rests, and sake cups for the customers to choose from. All of the sake is from the owner-chef's native Fukui prefecture. It matches well with Japanese cuisine, and synergizes with the food. Some are seasonally limited or rarely sold in Tokyo, or from small breweries. There are also natural wines and homemade fruit liqueur. The atmosphere allows the body and soul to unwind, and the chance to have a peaceful moment with good sake.

(H) Tue-Sat 18:00-23:00 (L.C.22:30), Sun 17:00-22:00 (L.C.21:30) (T) 03-6804-9980 (C) Mon
(Ad) 1-1-20 Uehara, Shibuya-ku (Ac) 1- min. walk from Yoyogi Hachiman Station (Odakyu Odawara Line)
(U) ha-re-ma.com

涩谷的酒鬼横丁

在涩谷的酒鬼横丁
一家接一家地喝下去

Shibuya Nonbei Yokocho "Drinkers Alley"
Bar-hopping

第二章

涩谷的酒鬼横丁（涩谷）

"酒鬼横丁"是涩谷站旁电车轨道沿线附近一条充满昭和时代气息的居酒屋街，横丁在日语中指"小巷"。涩谷的酒鬼横丁内目前有38家居酒屋开门营业，其中"松菊""Chotto Yottette""Yasaiya"（蔬菜屋）这几家尤以清酒种类丰富著称。多数店铺面积狭小，所以在酒鬼横丁也有着独特的"酒鬼守则"。无论是只点一杯酒长时间赖着不走，还是跟同伴吵闹喧哗，在这里都不被允许，喝完赶紧给后面的人让地方才合规矩。所有店铺都欢迎新客人加入，所以看上哪家店直接推门进去就是。跟邻座陌生人边喝边聊也是酒鬼横丁的玩法之一。

(时)以各店为准　(电)各店不同　(休)多数店铺星期六、星期日休息　(址)涩谷区涩谷1-25-9、1-25-10　(交)JR涩谷站步行3分钟　(网)www.nonbei.tokyo

SHIBUYA NONBEI YOKOCHO (Shibuya)

Located along the railway track just a short walk from Shibuya station. "Nonbei Yokocho" is an izakaya town with the atmosphere of the Showa era. There are currently 38 shops. "Matsugiku," "Chotto Yottette," and "Yasaiya" have plenty of varieties of Japanese sake. All of the shops are small, so it is frowned upon to stay for a long time ordering just 1 cup of sake or to make lots of noise with a group of friends. The unspoken rule is to give your seat to the next customer when your drink is finished. All shops welcome new visitors, so please feel free to step into any of the shops. Talking with the person sitting next to you is one of the charms of Nonbei Yokocho.

(H) Depends on the shop　(C) Most bars are closed Sat, Sun　(Ad) 1-25-9,10 Shibuya, Shibuya-ku　(Ac) 3-min. walk from Shibuya Station (JR Line)　(U) www.nonbei.tokyo

渋谷的酒鬼橫丁

大多数酒鬼横丁内的居酒屋在昭和时代都只是路边摊，所以面积都不大，几个人就可以把店内坐满。因此去时不呼朋引伴，坐下后时时顾虑到身边其他客人，边喝边与陌生客人聊天才符合这里的规矩。"松菊"的老板娘每日轮换，店内上下全靠老板娘一人照应，不同老板娘当值时店内氛围完全不同。店内通常供应 4 至 6 种清酒。到了冬天，热乎乎的关东煮最受欢迎。

Most bars in Nonbei Yokocho become quickly packed as they are very small, just as they had started as street vendors in the Showa era. It is important not to go in a large group, to be respectful of other customers, and talking with the other customers is one of the appealing factors. "Matsukiku" is run by one female bar manager and always has about 4~6 types of sake. *Oden* (Japanese hotpot dish) is a popular dish during winter.

从涩谷站八公口算起的话，"Chotto Yottette"位于酒鬼横丁的最深处。老板娘开朗爽利，店内供应的地方酒种类丰富。该店还有东京少见的东出云地区产的"王禄"酒，不少酒客都是专为它而来。

The last bar when going from Shibuya station's Hachiko exit is "Chotto Yottette." The bar has a friendly female bar manager and there are many types of local sake. Many people visit this bar to drink the sake "Ouroku" brewed in the town of Higashi Izumo, which is rarely found in the city.

"Yasaiya"（蔬菜屋）店如其名，主要提供蔬菜类佐酒料理。店内供应的清酒全部经过店主精挑细选。二楼设有座席。由于该店人气颇高，建议事先预约。与横丁内的另一家酒馆"APRE"是姊妹店。

"Yasaiya," as the name suggests, is a bar that serves vegetable otsumami and offers a unique lineup of sake. There are tatami floor seats on the 2nd floor. It is popular and a reservation is recommended. Bar "APRE" is operated by the same owner.

SHIBUYA NONBEI YOKOCHO

第三章

学习日本酒的相关知识

LEARNING ABOUT SAKE

养成习惯，
记住自己喜欢的酒的品牌和口味

过去年轻人不怎么喝清酒。"2011年东日本大地震后，为了复兴受灾的东北地区的经济，越来越多的人开始购买东北地区生产的清酒，对清酒本身的关注度也随之提高。"日本酒信息馆馆长今田周三介绍道。而且，如今越来越多在外学习了科学制酒技术的年轻人回到自己的家乡酒藏，"他们按自己的口味喜好酿造的新一代清酒改变了传统清酒陈旧过时的形象"。对于那些平常没有饮酒习惯、不知道该喝什么酒好的人，今田建议"一是可以在买酒时问问店家的意见，二是在遇到自己喜欢的酒时，要记住它的品牌、口味和种类，这样下次选酒时就有了基准"。

第三章

Learn to Remember the Brand and Flavor
of Sake that You Tasted and Liked

Sake was not really enjoyed by the younger generation until some years ago. "After the 2011 Tohoku earthquake and tsunami, more people started buying sake produced in Tohoku as a way of encouraging the people from there," says Shuso Imada, director of the Japan Sake and Shochu Information Center. Now, young people tend to come back to breweries after studying sake from a scientific approach, and "are changing the unromantic image of sake by creating their own style that they are interested in drinking." People who do not drink sake regularly find it hard to decide what to purchase, but "asking the store is one method, and remembering the brand, flavor, and type of sake when you taste something you like is also a good basis for selecting your next sake."

ARTRIP ADVISER
艺术之旅顾问

今田周三
Shuso Imada

日本酒信息馆馆长，在任3年。极爱清酒。"配合料理饮用清酒时，清酒的味道会随料理而变。所以要想充分享受清酒美味，一定要在食材搭配上多下功夫。"

Director of the Japan Sake and Shochu Information Center. He has been the director for 3 years and is a fan of sake. "When having sake with food, the taste of sake changes depending on the accompanying food. I recommend being careful with the pairing in order to enjoy sake."

⓯ 日本酒信息馆（虎之门）

2016年重装开放的日本酒信息馆由联合了日本各地清酒、烧酒酒藏的全国工会组织——日本酒造组合中央会运营，这里可谓是一处广泛收集各地酒藏及清酒信息的"情报站"。日本酒造组合中央会的主要活动目的是向海内外传播清酒、烧酒、泡盛酒、味淋（甜料酒）等日本酒的魅力，馆内提供英文版、中文版、韩文版的平板终端导览服务。最让人兴奋的是，馆内提供100余种清酒及本格烧酒的试饮，许多上班族选择下班后来这里喝上一杯再回家。墙上展示的当季清酒每三周一换。"清酒的学问实在是太深了，不光是其出产的酒藏，包括搭配进食的料理、饮用方法等在内的许多因素都会改变酒的味道。我们希望可以有更多的朋友轻松随意地来这里获取有关各种日本酒的信息。"

㈠10:00—18:00 ㈢03-3519-2091 ㈤星期六、星期日、节假日、年末年初 ㈥港区西新桥1-6-15日本酒造虎之门大厦1层 ㈦地铁虎之门站步行3分钟 ㈧www.japansake.or.jp

⓯ JSS INFORMATION CENTER (Toranomon)

The JSS was renovated and reopened in 2016. The center is run by a group of sake and shochu breweries, and therefore has a wide range of information on breweries and sake from all over Japan. The aim of JSS is to provide information on sake, shochu, *awamori* (distilled alcoholic beverage indigenous to Okinawa, Japan), *mirin* (sweet rice wine used for cooking) to Japanese people and to many foreigners, and offers tours using tablets in English, Chinese, and Korean. The center offers sake tasting and always has a pleasing number of 100 brands of sakeand shochu arranged by type. Many people also visit to enjoy a quick drink after work.The exhibition area along the wall showcases the sake in season, and the lineup changes every 3 weeks. According to JSS, "Sake is a beverage of many depths in which the taste varies according to the brewery, depending on the food pairing, and different ways of drinking it. We hope our center can be a place where visitors can drop by to get information on various old and new sake."

(H) 10:00-18:00 (T) 03-3519-2091 (C) Sat, Sun, National holidays, Year-end holidays (Ad) Nihon Shuzo Toranomon Bldg. 1st floor, 1-6-15 Nishishinbashi, Minato–ku (Ac) 3-min. walk from Toranomon Station (Tokyo Metro Toranomon Line) (U) www.japansake.or.jp

信息馆内装饰的桶装清酒——樽酒。清酒通常以300毫升、720毫升、1.8升的规格贩卖，但在进奉神社或结婚典礼、庆祝活动中举行"开镜"仪式时会使用樽酒。樽酒一般分为18升、36升、72升三种规格。

A *taruzake* (sake stored in a cask) displayed at JSS. Sake is usually sold in 300 ml, 720 ml, or 1.8 l bottles, but is contained in casks when given as offerings at shrines or used for "*Kagami Biraki*" (a Japanese tradition in which a cask of sake is opened at the beginning of a ceremony) at weddings or other celebrations. Taruzake comes in 3 standard sizes of 18 l, 36 l, and 72 l.

什么是清酒？
WHAT IS SAKE？

以大米为原料发酵而成的日本传统酒饮

清酒同本格烧酒被称为日本"国酒"。作为历史悠久的日本传统酒饮，清酒"身影"也经常出现在各种国际性宴席上。关于清酒的起源众说纷纭，《三国志·魏书·倭人传》中关于酒的相关记载表明日本在很久以前便已开始制酒。酒在传统祭祀活动中同样不可或缺，日本自古就有向神社供奉"御神酒"的传统。清酒由大米、酒曲和水发酵而成，与葡萄酒、啤酒同属酿造酒，只不过制造工艺更为精细复杂。由于清酒原料只有大米、酒曲和水这三样，原料处理方法、酿酒技术以及产地气候则成了左右成品品质的关键要素。

It is an alcoholic beverage made by fermenting rice that the people of Japan started drinking from long ago

Sake and traditional shochu, together, were referred to as "Kokushu", and were served in domestic and international events as traditional Japanese alcoholic beverages. Although the origin of sake is unclear, it is believed to have existed from long ago, as it appears to be described in the book *Gishi-wajinden* (written c. 280). Sake has also been an essential part of holy rituals from early times and offered at shrines as *Omiki* (sake offered to gods). It is made by fermenting rice, *koji* (rice malt), and water, and is categorized with wine and beer due to the similar brewing process. However, its brewing process is more complex and delicate than the other 2 beverages. As it utilizes only rice, koji, and water, the finished product is completely different depending on the handling of each ingredient as well as other factors including skills and climate.

清酒的种类
TYPES OF SAKE

清酒依原料大米精磨程度分为"特定名称酒"和"普通酒"。"特定名称酒"又分两种,一种是只用水、大米、酒曲制成的纯米酒,另一种则会在纯米酒中加入少量酿造酒精进行调味。

Sake is classified as *"Tokutei-meisho-shu"* (premium sake) or *"Futsu-shu"* (ordinary sake) depending on the rice polishing ratio. Tokutei-meisho-shu is further divided into *Junmai* (pure rice sake), which is made of water, rice, and koji, and sake made by adding a small amount of brewer's alcohol to Junmai for added flavor.

纯米大吟酿
JUNMAI DAI GINJO

吟之舞

东京都，田村酒造场，2484日元（720毫升）。香气华丽，味道精纯。东京峰会用酒。

GINNOMAI

Tokyo/Tamura Shuzojou / ¥2,484 (720ml). Fragrant aroma and refined umami. It was served at the Tokyo Summit.

繁桝

福冈县，高桥商店，7560日元（1800毫升）。使用"酒米之王"山田锦大米为原料。兼具纯米清酒的口感与大吟酿的华丽香气。

SHIGEMASU

Fukuoka prefecture/ Takahashi Shoten/ ¥7,560 (1800ml). Made with Yamada Nishiki rice. Tastes of Junmai and has the fragrance of *Dai Ginjo* (super premium sake).

纯米大吟酿只用水、大米和酒曲制成，将大米的美味发挥至极致是其主要特征。不仅如此，纯米大吟酿只选用精米步合度在50%以下的大米做原料，被磨去外层所剩下的米芯经低温长期发酵而形成的酒浆口感醇正，香气华丽，代表了日本清酒的最高品质。

Junmai-shu is sake made of water, rice, and koji. The unique point of this sake is that it brings out the natural umami of rice. Furthermore, Junmai Dai Ginjo-shu is specially made by using pure rice that has been polished to 50% or lower. The pure center of the rice is fermented for a long time at a low temperature, producing a pure flavor and a fragrant aroma of Ginjo-shu, making it the most premium sake.

荣四郎

福岛县，荣川酒造，10800日元（1800毫升）。使用日本"百选名水"之一的高级泉水。香气优雅，甘味鲜爽。

EISHIRO

Fukushima Prefecture/ Eisen Shuzo/ ¥10,800 (1800ml). It is made using one of the 100 best waters of Japan. Elegant aroma and crisp sweetness.

富士正

静冈县，富士正酒造，2970日元（720毫升）。吟酿香气高雅不凡，充满果香。口感微干，味道清爽。

FUJIMASA

Shizuoka Prefecture/ Fujimasa Shuzo/ ¥2,970 (720㎖). Refined and fruity Ginjo aroma. Slightly dry and crisp flavor.

大吟酿
DAI GINJO

纯米大吟酿由水、大米和酒曲酿造而成。在纯米大吟酿中添加少量酿造酒精而酿成的酒被称为大吟酿。因其精米步合度在50%以下，所以蛋白质及脂质的含量低，口感醇正。与纯米大吟酿相比，香气更华丽，口感更爽口。

Dai Ginjo is made by adding a small amount of brewer's alcohol to Junmai Dai Ginjo, which is sake made of water, rice, and koji. The rice used to make Dai Ginjo is polished to 50% or lower, taking away the protein and fat, both of which have a bitter taste, thereby refining the flavor. More fragrant and crispy than Junmai Dai Ginjo.

菱正宗

广岛县，久保田酒造，5832日元（1800毫升）。芳醇华丽，入口醇和。

HISHI MASAMUNE

Hiroshima Prefecture/ Kubota Shuzo/ ¥5,832 (1800㎖). Fragrant rich aroma and mild flavor.

北之庄

福井县，舟木酒造，5400日元（1800毫升）。使用精磨至35%的山田锦酒米为原料。口感清爽，入喉柔顺。

KITANOSHO

Fukui Prefecture/Funaki Sake Brewery/ ¥5,400 (1800㎖). Yamada Nishiki rice polished to 35%. Very fresh and smooth.

武藏之里

冈山县，田中酒造场，4104日元（1800毫升）。使用美作当地产的山田锦酒米。具有柔和的香气及味道。

MUSASHI NO SATO

Okayama Prefecture/ Tanaka Shuzojou/ ¥4,104 (1800㎖). Made using Yamada Nishiki produced locally in Mimasaka city. Soft aroma and umami.

纯米吟酿
JUNMAI GINJO

第三章

纯米吟酿只用水、大米和酒曲制成，精米步合度在60%以下。由于精米步合度略高于大吟酿，各款纯米吟酿间口感差异较大，可以较为明显地感受到大米本味的馥郁柔和。此种酒香气华丽，定价实惠，相比其他种类清酒人气更高。

Junmai Ginjo-shu is sake made of water, rice, and koji, and is polished to 60% or less. Since the rice polishing ratio is not as low as Dai Ginjo, some of Junmai Ginjo have a larger range of flavors and have the soft umami of fuller and less polished rice. It has a fragrant Ginjo aroma and is available at a reasonable price, which makes it more popular than other sake.

隐岐誉

岛根县，隐岐酒造，2847日元（1800毫升）。爽快的酸味与鲜味完美调和，温冷皆宜，回味悠长。

OKIHOMARE

Shimane Prefecture/ Okishuzo/¥2,847 (1800㎖). It has a good balance of crisp acidity, umami and depth, served at room temperature or as warmed sake.

73

出羽樱（樱花）

山形县，出羽樱酒造，2808日元（1800毫升）。凭借醇厚果香及馥郁米香，自发售以来屡获大奖。

DEWAZAKURA (OKA)

Yamagata Prefecture/ Dewazakura Sake Brewery/¥2,808 (1,800㎖). Fruity with full flavor. Winner of numerous awards since its launch.

吟酿
GINJO

在纯米吟酿中添加少量酿造酒精而制成的酒。由于精米步合度较低，在60%以下，在清酒特有的吟酿果香中又富含大米的馥郁口感。与口感浓厚的纯米吟酿相比，吟酿的味道更淡雅清丽，易于入口。

A sake made by adding a small amount of brewer's alcohol to Junmai Ginjo made of water, rice, and koji. It has a rice polishing ratio of 60% or lower, giving it the fruity aroma of Ginjo, while keeping the well-rounded flavor of rice. Compared to the robust and rich Junmai Ginjo, Ginjo is crisper and lighter.

曾我之誉

神奈川县，石井酿造，3456日元（1800毫升）。口味淡雅清丽，口感偏干。香气新鲜轻快，味道馥郁。

SOGA NO HOMARE

Kanagawa Prefecture/Ishii Jozo/ ¥3,456 (1800ml). Clean and dry. Retronasal aroma is fresh and light with a refreshing flavor.

南部美人（特别纯米酒）

岩手县，南部美人，2916日元（1800毫升）。使用自家培育的酒米"Gin-otome"。适合配餐饮用。

NANBUBIJIN
(Tokubetsu Junmai-shu)

Iwate Prefecture/Nanbu Bijin/¥2,916 (1800㎖). It is made using homemade "Gin-otome" sake rice. It goes well with any food.

纯米酒
JUNMAI-SHU

只用水、大米、酒曲制成的酒。过去曾有规定，纯米酒的精米步合度必须在70%以下，现已取消。有些制造纯米酒的酒造为了更大程度地激发米香，精米步合度甚至会达到80%～90%。因此不同纯米酒的口感从淡丽到浓厚，跨度很大。香气色泽皆出众的纯米酒又被称为"特别纯米酒"。

A sake made of water, rice, and koji. The rule of rice polishing ratio had always been established at 70% or lower, but there are currently no rules. Some sake have a ratio as high as 80 to 90% in order to bring out the unique flavor of rice. Therefore the flavor ranges from clean to robust. Sake with good quality aroma and color is known as "Tokubetsu Junmai-shu."

七田（纯米酒）

佐贺县，天山酒造，2592日元（1800毫升）。米香四溢，味道近似酸味平衡良好的香槟。

SHICHIDA
(Junmai-shu)

Saga Prefecture/Tenzan Sake Brewer/¥2,592 (1800㎖). Umami of rice and flavor of champagne with balanced acidity.

第三章

北之锦（纯米酒）

北海道，小林酒造，2376日元（1800毫升）。轻快爽口。因未经过滤，酒体呈清酒原本的稻穗色。

KITA NO NISHIKI
(Junmai-shu)

Hokkaido Prefecture/ Kobayashi Shuzo/¥2,376 (1800㎖). Light and crisp. It is unfiltered and a gold-colored sake.

久住千羽鹤（纯米酒）

大分县，佐藤酒造，1150日元（720毫升）。味道浓醇，酸味强烈。酒名由文豪川端康成所取。

KUJU SENBAZURU
(Junmai-shu)

Oita Prefecture/Satoh Distilling LLC/¥1,150 (720㎖). Clean and delicate with acidity. Branded by writer Yasunari Kawabata.

本醸造
HONJOZO

三千盛

岐阜县，三千盛，2270日元（1800毫升）。甜味、酸味、涩味完美平衡，口感偏干。适合搭配一切和食。

MICHISAKARI

Gifu Prefecture/Michisakari/¥2,270 (1800 ml). Dry sake with a good balance of sweetness, acidity, and astringency. Suits any Japanese food.

超特三千盛

岐阜县，三千盛，2780日元（1800毫升）。口感纯粹，无多余杂味，米香明显。

CHOTOKU MICHISAKARI

Gifu Prefecture/Michisakari/¥2,780 (1800ml). Clear without any extra bitterness and has the proper umami of rice.

本酿造是用水、大米、酒曲，再添加少量酿造酒精而制成的酒。精米步合度在70%以下，按规定每1吨大米可添加不多于100千克的酿造酒精。极其爽口，适合搭配各类料理。"特别本酿造"指香气色泽皆出众的精品本酿造。

Honjozo-shu is made of water, rice, and koji, with a small amount of brewer's alcohol added. The rice polishing ratio is 70% or lower, and the rule is that a maximum of 100kg of brewer's alcohol can be added per 1 ton of rice. It is uniquely crisp and can be easily paired with various dishes. "Tokubetsu Honjozo-shu" refers to sake with the best aroma and color out of the Honjozo.

江田岛

广岛县，江田岛铭酿，1657日元（720毫升）。入口清爽，回味丰富，口感偏干。

ETAJIMA

Hiroshima Prefecture/Etajimameijyo/ ¥1,657 (720ml). Fresh but maintains richness. A little bit on the dry side.

普通酒
FUTSU-SHU

黑松剑菱

兵库县，剑菱酒造，2 533 日元（1800 毫升）。浓厚香味及浓醇米味在入口瞬间即扩散开来。

KUROMATSU KENBISHI

Hyogo Prefecture/Kenbishi Sake Brewing/ ¥2,533 (1800ml). Quickly spreads a rich aroma and the clean and delicate flavor of rice.

普通酒指纯米酒、本酿造等特定名称酒之外的清酒。有的使用精米步合度 70% 以上的大米，有的还使用水、大米、酒曲、酿造酒精以外的其他原料，有的酿造酒精量超过 10%，种类繁多。味道变化多端，百饮不厌，价格亲民，是每晚小酌一杯的好选择。

Futsu-shu refers to pure sake that is not made under the standards required for Tokutei-meisho-shu such as Junmai-shu and Honjozo-shu. Some are made using rice with a polished ratio of 70% or more, while some are made of ingredients other than water, rice, koji and brewer's alcohol, and others contain more than 10% brewer's alcohol. The flavors vary and are available at a friendly price, making them good everyday dinner sake.

生酒
NAMAZAKE

WAKAMUSUME

山口县，新谷酒造，（右）2160日元(720毫升),（左）1620日元(720毫升)。香气收敛，充满大米原本的柔和甜味。

WAKAMUSUME

Yamaguchi Prefecture/ Shintani Shuzo/(R) ¥ 2,160 (L) ¥1,620 (720ml) each. Gentle aroma, and the gentle natural sweetness of rice.

左大臣 粹 (SUI)

群马县，大利根酒造，3024日元（1800毫升）。香气温和，甜味适中。属于未经调整的纯粹生酒，可加冰块饮用。

SADAIJIN SUI

Gunma Prefecture/ Otone Sake Brewing/ ¥3,024 (1800ml). Mildly aromatic and sweet. Pure sake that can be enjoyed on the rocks.

制酒时，一般在过滤后会进行名为"过火"的加热处理，在制作工程中从未"过火"的酒被称为生酒。生酒种类丰富，有"纯米生酒""吟酿生酒""本生"等。一般需冷藏保存，因品质较难管理，只有在酒造才能品尝到刚刚做好的新鲜生酒。与"生贮藏酒""生诘酒"是不同种类。

While sake is usually pasteurized, after the pressing process, in a heating process called hi-ire, Namazake is never heated. Types of Namazake include "Junmai Namazake" and "Ginjo Namazake," and it is also known as "Hon-nama." It usually requires refrigeration and is hard to maintain its quality, but can be enjoyed fresh at breweries. It is not the same type of sake as "*Namachozo-shu*" (sake pasteurized prior to bottling) or "*Namazume-shu*" (sake pasteurized prior to storing).

天览山 本生

埼玉县，五十岚酒造，1101日元(720毫升)。果香充盈，入口轻快。

TENRANZAN HONNAMA

Saitama Prefecture/ Igarashi Shuzo/ ¥1,101 (720ml). Fruity aroma of Namazake.

有机
ORGANIC

千代 MUSUBI

鸟取县，千代 Musubi 酒造，1890 日元（720 毫升）。有机农作物加工酒类认证。果香充沛，口感轻盈。

CHIYO MUSUBI

Tottori Prefecture/Chiyo Musubi/¥1,890 (720ml). Approved as an organically processed alcoholic beverage. Fruity and light.

一之藏有机纯米酿造

宫城县，一之藏，3186 日元（1800 毫升）。用有机米酿造的特别纯米酒（限定产品）。易入口，味道馥郁。

ICHINOKURA YUUKIJUNMAI-JIKOMI

Miyagi Prefecture/Ichinokura/¥3,186 (1800 ml). It is a Junmai-shu made of organic rice (limited product). Easy to drink with full flavor.

近几年，使用有机栽培葡萄酿造的葡萄酒颇受关注，在年轻一代中很受欢迎。受其影响，越来越多的酒造开始用种植过程中未使用农药及化学肥料的大米酿造有机清酒。有机大米的香甜之气，再加上柔和的口感是其主要特征。有机清酒市场有望在未来几年持续增长。

Recently, natural wines (bio-wines) made with organically grown grapes are drawing more attention and especially popular among the younger generation. This fad has caused an increase in breweries that make organic sake using pesticide and fertilizer-free rice. The umami, sweetness, and gentle taste of organic rice is unique to this sake. This market is expected to continue growing.

菊水纯米吟酿有机清酒

新潟县，菊水酒造，695 日元（300 毫升）。100% 使用新潟县产有机米。香气宜人，口感清爽。

KIKUSUI ORGANIC JUNMAI GINJO

Niigata Prefecture/Kikusui Sake Co./¥695 (300ml). Brewed from 100% organic rice from Niigata. Fragrant and clean.

其他
ETC.

古酒
KOSHU

陈酿 3 年以上的清酒称"古酒"或"熟成酒",抑或"熟成古酒"。新酒不具备的复杂香味及柔和口感是古酒的最大魅力。

Sake that has been aged for 3 years or more is called "*Koshu*", "*Jukusei-shu*" or "*Jukusei-koshu.*" They have complex flavors with a smooth finish that cannot be tasted in young sake.

熟露枯

栃木县,岛崎酒造,5 406 日元(720 毫升)。在酒藏的洞窟内经 5 年陈酿的大吟酿酒。香气幽深。

UROKO

Tochigi Prefecture/ Shimazaki Shuzo/ ¥5,406 (720ml). Dai Ginjo aged for 5 years or more in a cave owned by the brewery. Deep aroma.

十二六 LIGHT

长野县,武重本家酒造,1234 日元(720 毫升)。酸甜适宜,味道清爽。

DOBUROKU LIGHT

Nagano Prefecture/ Takeshige Honke Brewing Corp./ ¥1,234 (720ml). Not too sweet, and crisp acidity. Light smooth flavor.

浊酒
NIGORI

原浆压榨后只经过轻微过滤的酒。酒体呈混浊的白色,甜味浓郁,未经过火,具较强起泡性。

Sake made by filtering the fermented rice particles using a broader mesh. It is white and cloudy and rich in sweetness, is unpasteurized, and bubbly.

七贤 山之霞

山梨县，山梨铭酿，1944 日元（720 毫升）。新鲜酵母直接装瓶而成，吟酿香气偏果香。

SHICHIKEN YAMANO KASUMI

Yamanashi Prefecture/ Yamanashi Meijo/ ¥1,944 (720ml). Bottled natural yeast. Strong aroma of Ginjo-shu with fruitiness.

⟨辛口⟩ DRY

气泡清酒
SPARKLING

颇受女性欢迎的一种清酒。气泡清酒酿造法主要分为注入二氧化碳法及瓶内二次发酵法两类。

Popular among women. There are 2 types of sparkling sake. One type is carbonated by adding carbon dioxide, and the other by secondary bottle fermentation.

⟨甘口⟩ SWEET

NENE

山口县，酒井酒造，756 日元（300 毫升）。采用瓶内二次发酵法制成，拥有温和的起泡感。甜味清幽，酸味爽口。

NENE

Yamaguchi Prefecture/ Sakai Shuzo/ ¥756 (300ml). Fermented in the bottle, and slightly bubbly. Mildly sweet and crisp acidity.

第三章

贵酿酒
KIJOU-SHU

在酿造过程中，使用清酒代替部分作为原料的水，再加入大米和酒曲酿造而成的酒。甜味浓厚，口感黏稠。

Sake made by partially using sake instead of water. Presents a rich sweetness and melts in the mouth.

满寿泉

富山县，升田酒造，2376 日元（500 毫升）。熟成后酒体呈深琥珀色。属高级清酒，类似葡萄酒中的贵腐酒。

MASUIZUMI

Toyama Prefecture/ Masuda Shuzo/ ¥2,376 (500ml). Becomes a rich golden-brown when aged. Very similar to *Kifu wine* (botrytized wine).

解读清酒标签
HOW TO READ SAKE LABELS

标签可谓是清酒的"脸面"。虽然清酒标签上标注的内容是固定的,但由于没有明确规定具体的标示方法,不同酿酒商的标签会有些微差异。以下三个例子将帮助你学习清酒基本的标签布局及解读方法,从而加深对每款酒的理解。

Labels, it can be said, are the "face" of sake. All sake products are obliged to have labels indicating certain matters, but since there are no strict rules, the labels differ slightly depending on the manufacturer. Here are some samples for you to understand the layout of the labels and profile of each brands.

例 1

左图标签中最上面一行的前两个字是酒名,后四个字指酿造方法,"纯米吟酿"是酒的种类,其下是酒造提供的解说及建议饮用方法。再往下的"清酒"表明该酒满足日本酒税法所规定的酿造条件,属必须标示的内容之一;清酒有时也可标示为"日本酒"。其他必须在酒标内进行标示的还有酒造名称、地址、原材料、酒精浓度、容量、生产日期这6项。照片中的酒在标签最下方标示了酒造名称及生产日期。

The word Sohomare at the very top is the name of the sake, and namajikomi is the manufacturing method. Junmai Ginjo is the type of sake, and is followed by the brewer's explanation on the sake and how to drink it. Seishu is also labeled as "Sake" and must indicate that it has satisfied all manufacturing requirements under the Liquor Tax Act. It must also indicate 6 other items including the name and address of the brewer, ingredients, alcohol - level, volume, and date of manufacture. The sake bottle in this photograph indicates the name of the brewer, date of manufacture, and other information at the bottom.

例 2

纵向排版的酒标。关于酒标应该横向排版还是纵向排版,并无明确规定。只要必须标注的内容没有遗漏,酒标的排版设计及文字表现的自由度其实很高。照片中的酒标包含该酒的历史背景介绍及一段饱含酒造酿酒热忱的文字。从酒标设计中可以看出一家酒造的个性及酿酒的理念。

This label is written vertically. There are no rules on whether labels should be horizontally or vertically written. As long as they clearly indicate the required information, there are no regulations for design or other indicated information. The sake in this photo shows the history of how it was created, and a warm message from the brewer. The designs on sake labels showcase the uniqueness of each brewer, their passion, and their policies.

第三章

例 3

用有机大米酿造的"有机"清酒的标签。除酒名、酒造名等必须包含的内容外,还特别标注使用有机农作物酿造,方便消费者选购。至于"日本酒度"的标识,过去一般认为正值代表辛口,负值代表甘口,不过由于现在清酒口感味道愈发丰富,已不能一概而论。

This is a label for organic sake manufactured using organically produced rice. The label must indicate the names of the product and the brewer, and must also indicate proof that the sake was manufactured using organic products in order to make selection easier. The label also shows the degree of the flavor, where a "+" indicates that is dry and a "−" indicates sweetness, but since the variety in flavors of sake in recent years has become incredibly vast, it is often quite hard to categorize their flavors.

挑选清酒酒器
CHOOSING A SAKE SET

品尝清酒时酒温固然重要,酒器同样会影响酒的味道。酒器种类繁多,在条件允许的情况下选择合适的酒器有助于最大程度上展现酒的风味。

The taste of sake varies with temperature, but the sake set also influences the flavor. Choose the most suited set among the varieties of sake sets and savor the unique flavor.

德利和猪口

德利(日本酒酒壶)一般分1合及2合两种规格,其形状大小基本决定了盛装酒量的多少。有时会用带单侧导流口的"片口"代替德利。根据形状,猪口(日本酒酒盅)有时也被称为"盃"(盘盏状酒器)或"Guinomi"(形状与猪口相近,尺寸大于猪口)。德利和猪口主要用于饮用温酒。

TOKKURI AND OCHOKO

Tokkuri (a ceramic flask) is traditionally used for serving sake in units of one *go* or two *go* (180 ml or 360 ml). *Katakuchi* (a lipped flask without a handle) is also used instead of tokkuri. *Ochoko* (a small drinking cup) is called "sakazuki" and "guinomi" according to the shape. This small cup is perfect for drinking heated sake.

玻璃酒器

猪口有各种颜色、形状及材质,在饮用冷酒等冰镇后口感更佳的清酒时推荐使用玻璃猪口。除猪口这类小号酒器外,还有尺寸更大的玻璃酒器,选择丰富。

GLASS SAKE CUP

There are different colors, shapes and materials of ochoko but a glass cup is perfect for drinking cold sake and sake with a refreshing taste such as *reishu* (chilled sake). There are varieties of glass cups from ochoko-size to large-size.

葡萄酒杯

葡萄酒杯适合像葡萄酒一样香气充沛的清酒。使用葡萄酒杯饮用纯米大吟酿或纯米吟酿,可以更好地品鉴出它们的清爽口感与馥郁酒香。近来市面上甚至出现了专为饮用清酒而制造的葡萄酒杯。

WINE GLASS

Wine glasses are perfect for aroma-rich sake like wine. You can savor the fresh and fruity aroma such as Junmai Dai Ginjo-shu and Junmai Ginjo-shu. Wine glasses designed especially for sake are available nowadays.

第三章

玻璃杯 + 升

常见于清酒专卖店的一个酒器组合。在使用"玻璃杯+升"这套酒具饮酒的时候,倒酒一般会倒至酒从玻璃杯中满溢出来,这种饮酒方法也被称为"mokkiri"。如果客人下一杯点的还是同一款酒的话,店家可能会让其继续使用同一套酒器。

GLASS + MASU

A set of glass and *masu* (a wooden drinking box) are commonly used at sake specialized stores. Traditionally, sake is poured into a glass placed in a masu until it overflows from the rim of the glass. This is called "mokkiri". If you ask for the same brand of sake for a refill, it may be poured into the same glass and masu.

升

分为用漆器制作的升及木升两种,前者历史更悠久,原本为量酒工具。昭和三十年代(1955年左右)升开始作为酒器使用,木升也变得更受大众欢迎。

MASU

Lacquer masu has an older history than the currently available wooden masu, which were originally used for measuring an amount of sake. People started to use masu as sake cups in around 1955, then the wooden masu became more popular and widely used now.

品酒专用的蛇目猪口
BULLSEYE DESIGNED OCHOKO FOR SAKE TASTING

猪口底部常见的蛇目图案其实是为了判断酒的品质好坏而特别设计。白色部分可用来检查清酒的"透明度",蓝色部分则用来检查清酒的"光泽度"。

You can often see a bullseye in the bottom of ochoko. This is originally used for sake tasting which determines good or bad sake. The white part of the bullseye measures the "transparency" and the blue part tells you the "glossiness."

去青梅的小泽酒造
来一次清酒主题一日游

小泽酒造（青梅）

从东京都中心乘一小时左右的电车到达奥多摩地区的泽井站，下车后步行 5 分钟，就到了小泽酒造。开业于 1702 年的小泽酒造如今已有三百多年的历史，其最为出名的产品当属东京地区代表性地方酒"泽乃井"。在这里，游客可以参观到江户时代（1603—1867）建造的酒藏及汲取酿酒用水的涌井，感受酒造特有的风情。想让这场酒造探访之旅更丰富愉快的话，可以顺路去一下酒造附近的日式庭院、美术馆及餐厅。参观需提前在小泽酒造官网预约。

(时) 10:00—17:00 (电) 0428-78-8210 (休) 星期一（遇节假日顺延） (址) 青梅市泽井 2-770 (交) JR 泽井站步行 5 分钟 (网) sawanoi-sake.com

Sake-themed Day Trip to
Sake brewery, Ozawa Sake Brewery in Oume

OZAWA SAKE BREWERY (Oume)

To get to Ozawa Sake Brewery, which is well known as a sake brewery of the popular "Sawanoi" brand in Tokyo, it takes about one hour on the Chuo line train from Shinjuku station, with a transfer to Oume line at Tachikawa station, followed by a 5 minute-walk from Sawai station in Okutama. Founded in 1702, this brewery with over 300 years of history has a sake warehouse which was built in the Edo period and well, which you can take a look at on a tour. Visiting the adjoining garden, art museum and restaurant would make for an exciting day trip. Reservation required for the tour.

(H) 10:00-17:00 (T) 0428-78-8210 (C) Mon (the following day if Monday is a National holiday)
(Ad) 2-770 Sawai, Oume. Tokyo (Ac) 5-min. walk from Sawai Station (JR Line) (U) sawanoi-sake.com

悬吊在房檐下的杉叶球叫作"杉玉",当绿色的杉玉被高悬于房梁之下时则代表有新酒酿成。图中建筑主要用来酿造及储藏清酒,小泽酒造内共有三栋此类建筑,其中一栋建于江户时代。

This round ornamental ball made of Japanese cedar hung under the eaves is called "sugitama". Green sugitama is a signal that *Shinshu* (new sake) is ready. The building at Ozawa Sake Brewery is used as the warehouse for the mashing and storing of sake. There are a total of three warehouses including the one built in the Edo period.

建于江户时代元禄年间(1688—1704)的储酒仓库造型别致,如今仍旧可以正常使用。土制墙壁可以为室内维持恒定的温度与湿度,让室内即使在炎热夏季也依旧凉爽;而稳定的温度也是对清酒贮藏起到至关重要作用的因素。该仓库目前主要用于酒类储藏。

The quaint warehouse built during the Genroku era (1688-1704) of the Edo period is still used. This earthern-wall warehouse is perfect for keeping the best condition of sake since it can keep the temperature constantly cool inside during summer. It is currently used as the warehouse for sake.

要想酿出好清酒,光有好的酿酒米还不够,好水同样不可或缺。小泽酒造酿酒用的水取自两处水源,其中一处名为"藏之井",其位置就在酒造后方,如今对外开放参观。

As well as high quality rice, delicious water also plays a very important role in brewing delicious sake. Water used at Ozawa Sake Brewery is from two particular water sources. One of them is the spring water from "the well of the storehouse" at the rear. This area is also available for tours.

与泡盛古酒一样,长期的贮存和熟成也会为清酒增添一丝别样味道。小泽酒造将放在仓库中贮藏的古酒命名为"藏守",只在特定时期对外出售。古酒是近来颇受关注的一个清酒种类。

Similar to koshu Awamori, the long aging process of sake also changes the flavor. The sake "Kuramori", also matured in a warehouse, is sold at Ozawa sake brewery for a limited time only. These koshu have gotten a lot of attention recently among other different types of sake.

OZAWA SAKE BREWERY

清酒的酿造过程
HOW IS SAKE MADE?

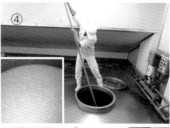

① 酿制清酒的原料之一的大米。小泽酒造只使用经过严格筛选的日本国产大米。

② 将在水中浸泡过的大米放在近似大型蒸锅的装置中蒸熟。如何根据季节调整水分及蒸米时间最考验酿酒师的功夫。

③ 在蒸好的米中加入酒母，制作酒曲。

④ 将发酵剂、酒曲、蒸米、水投入酿酒桶内，制作醪液。该步骤分三次进行，因此又称"三段酿造"。之后静置发酵 20 天左右。

⑤ 经压榨机提取的新鲜新酒。

⑥ 经过滤后残留的酒糟。

⑦ 贮存生酒"iroha"用的杉木桶。

① Rice, a key ingredient of sake making. Only carefully selected homegrown rice is used at Ozawa Sake Brewery. ② Steaming process of water-absorbed rice in a large steamer. Ensuring the right amount of water and the right timing in different seasons depends on the skills and experience of *toji* (a chief sake maker).
③ Add *tanekoji* (a yeast starter) to steamed rice to make *koji*. ④ Put *shubo* (a starter culture), koji, steamed rice and water into the barrel to make *moromi* (a fermentation mash.) The three different processes of moromi making are called "three-step preparation" and it takes 20 days for mash fermentation. ⑤ Freshly compressed new sake. ⑥ Left-over *Sakekasu* (sake lees) after the compression. ⑦ Cedar wooden barrel for storing of "Iroha".

在品酒吧台品鉴小泽酒造的名酒
SAVOR THE OZAWA SAKE BREWERY'S WELL-KNOWN BRANDS AT A SAKE TASTING BAR

第三章

小泽酒造除酒藏外还有一个面积不小的庭院,庭院内分布着商店和餐厅。餐厅内提供酒品的试饮(需付费),游客可以品尝到独家供应的当季新酿酒饮。右下角照片中是小泽酒造对外出售的原创酒杯。

Ozawa Sake Brewery has a big garden next to the warehouse, where shops and restaurants are dotted around the garden. Priced samples are offered, and you can savor the seasonal products served only here. Ozawa sake brewery's original guinomi (lower right) is available for purchase.

东京藏人

1800 日元(720 毫升)。使用精心培育的酒曲,采用传统酿造方法酿成的纯米吟酿。

TOKYO KURABITO

¥1,800 (720ml) Junmai Ginjo brewed with a traditional sake-brewing method.

特别纯米

1296 日元(720 毫升),2592日元(1800 毫升)。口感优雅,易入口。

TOKUBETSU JUNMAI

¥1,296 (720ml) ¥2,592 (1800ml) Smooth and sophisticated flavor.

木桶酿造 IROHA

1782 日元(720 毫升)。使用老木桶酿造的生酒,口感丰富而幽深。

KIOKEJIKOMI IROHA

¥1,782 (720ml) Brewed in a wooden barrel. Complicated deep flavor.

本酿造 SHIBORITATE

1102 日元(720 毫升)。拥有新酒特有的清新口感。季节限定。

HONJOZO SHIBORITATE

¥1,102 (720ml) Distinctive fresh taste of new sake. Seasonal item.

OZAWA SAKE BREWERY

东京都内和近郊酒藏名录
SAKE BREWERIES LIST AROUND TOKYO

武甲酒造（埼玉县秩父）

武甲酒造的代表酒是未在东京内流通出售的特别纯米酒"武甲正宗"。
酒造内设有免税店。可顺道游览相距不远的景点秩父神社。参观需事先预约。

(时) 8:00—17:30 (电) 0494-22-0046 (休) 无（年末年初除外）(址) 埼玉县秩父市宫侧町 21-27 (交) 秩父铁道秩父站步行 3 分钟 (网) www.bukou.co.jp

BUKOU SAKE BREWERY (Chichibu, Saitama)

Brewery famous for "Bukoku Masamune". Booking required.

(H) 8:00—17:30 (T) 0494-22-0046 (C) None(Except for Year-end and New Year holidays) (Ad) 21-27 Miyagawa-cho, Chichibu, Saitama (Ac) 3-min. walk from Chichibu Station (Chichibu-tetsudo Line) (U) www.bukou.co.jp

松冈酿造（埼玉县小川町）

始创于 1851 年（嘉永四年）。以"帝松"酒在清酒界闻名。松冈酿造会在每年 2 月
最后一个星期日为庆祝新酒酿成举行"帝松酒藏祭"。参观需事先预约。

(时) 9:30—17:00 (电) 0493-72-1234 (休) 年初 (址) 埼玉县比企郡小川町下古寺 7-2
(交) 在东武东上线小川町站附近乘鹰巴士至"北根"站，下车后步行 10 分钟
(网) www.mikadomatsu.com/index.html

MATSUOKA JOZO (Ogawamachi, Saitama)

Famous for "Mikadomatsu." Annual festival is held. Booking required.

(H) 9:30-17:00 (T) 0493-72-1234 (C) New Year holidays (Ad) 7-2 Shimofurutera Ogawamachi, Hiki-gun, Saitama (Ac) Ogawamachi Station (Tobu-Tojo Line) with a transfer to Eagle Bus and 10-min. walk from Kitane bus stop (U) www.mikadomatsu.com/index.html

丰岛屋酒造（东村山）

出产东京地方知名清酒"金婚正宗"和"屋守"。
每月对外举行一次夜间酒藏参观活动。参观需事先预约。

(时) 9:00—17:00 (电) 042-391-0601 (休) 星期六、星期日 (址) 东村山市久米川町 3-14-10
(交) 西武新宿线东村山站步行 15 分钟 (网) www.toshimayasyuzo.co.jp

TOSHIMAYA SAKE BREWERY
(Higashimurayama, Tokyo)

Brewery famous for "Kinkon Masamune" and "Okunokami." Booking required for the tour.

(H) 9:00-17:00 (T) 042-391-0601 (C) Sat & Sun (Ad) 3-14-10 Kumegawa-cho, Higashimurayama, Tokyo (Ac) 15-min. walk from Higashimurayama Station (Seibu-Shinjuku Line) (U) www.toshimayasyuzo.co.jp

东京都内和近郊酒藏名录
SAKE BREWERIES LIST AROUND TOKYO

石川酒造（福生）

酒造内拥有别致的白色土墙仓库和一间意式餐厅。
出产名酒为"多满自慢"。参观需事先预约。

(时) 各店铺不同 (电) 042-553-0100 (休) 各店铺不同 (址) 福生市熊川 1 丁目 (交) JR 拜岛站南口步行 20 分钟 (网) tamajiman.co.jp

ISHIKAWA SAKE BREWERY (Fussa, Tokyo)

Famous for "Tamajiman" and the Italian restaurant built on the premises.
Booking required for the tour.

(H) Varies depending on the facility (T) 042-553-0100 (C) Varies depending on the facility (Ad) 1 Kumagawa, Fussa, Tokyo (Ac) 20-min. walk from Haijima Station South Exit (JR Line) (U) tamajiman.co.jp

泷泽本店（千叶县成田）

位于成田山新胜寺参道。旗下"长命泉"酒是使用被誉为"百药之王"
的成田山水酿造而成。参观需事先预约。

(时) 10:30—16:00 (电) 0476-24-2292 (休) 星期六、星期日，节假日，年末年初 (址) 千叶县成田市上町 540 (交) JR 成田站步行 8 分钟 (网) www.chomeisen.jp

TAKIZAWA HONTEN SAKE BREWERY (Narita, Chiba)

Close to Naritasan Shinshoji Temple. Uses Naritasan spring water to brew "Chomeisen."
Booking required for the tour.

(H) 10:30-16:00 (T) 0476-24-2292 (C) Sat, Sun & National holidays, Year-end & New Year holidays (Ad) 540 Kamimachi, Narita, Chiba (Ac) 8-min. walk from Narita Station (JR Line) (U) www.chomeisen.jp

泉桥酒造（神奈川县海老名）

原料选用酒造自行栽培或签约栽培的神奈川县本地大米。只生产纯米酒。
出产名酒"IZUMI 桥"。参观（付费）需事先预约。

(时) 10:00—18:00 (电) 046-231-1338 (休) 星期日、节假日、年末年初 (址) 神奈川县海老名市下今泉 5-5-1 (交) JR 或相铁线或小田急线虾名站步行 20 分钟 (网) izumibashi.com

IZUMIBASHI SAKE BREWERY (Ebina, Kanagawa)

Uses only locally-grown rice for "Izumibashi" and Junmai-shu.
Booking required for the tour.

(H) 10:00-18:00 (T) 046-231-1338 (C) Sun & National holidays, Year-end & New Year holidays (Ad) 5-5-1 Shimoimaizumi, Ebina, Kanagawa (Ac) 20-min. walk from Ebina Station (JR Line, Sotetsu Line and Odakyu Line) (U) izumibashi.com

第三章

清酒小知识 2
清酒产地及其特征

北海道地区
出产的清酒口感清淡优雅。当地的酿酒史几乎与水稻种植史一样悠久。

东北地区
日本极为出名的几处大米产地之一,出产上好清酒的酒造也不少。该地清酒的口感特征一般被形容为清爽、强烈或浓厚。

北陆地区
比肩东北地区的名酒产区,好酒不在少数。新潟县和富山县出产的酒口感多偏清爽,石川县和福井县出产的酒口感较强劲。

关东地区
出产的清酒大多口感清爽。

中部地区
静冈县出产的酒清爽柔和。爱知县和岐阜县出产的酒口感多浓醇。

近畿地区
兵库县水质优良,同时也是被誉为"顶级酒米"的山田锦大米的主产地,故此地的酿酒业自古繁荣。该地出产的酒口感多厚实偏干。

中国地区
广岛县自明治末期起作为日本酒产地开始崭露头角,所产之酒口感多馥郁柔和。

四国地区
濑户内海一侧与太平洋一侧出产的清酒酒质完全不同。高知县出产的清酒口感清爽,爱媛县出产的清酒口味近似广岛清酒,入口馥郁柔和。

九州地区
作为著名烧酒产地,九州地区其实也有不少清酒酒藏。口感甘甜厚实是当地清酒的主要特征。

TRIVIA ON SAKE_2
SAKE PRODUCING REGIONS AND THEIR CHARACTERISTICS

HOKKAIDO
The sake features a light and delicate taste.

TOHOKU
Known as a leading rice producing region in Japan and many breweries that produce delicious sake are located in this region. The sake features a refreshing, strong or rich flavor.

HOKURIKU
Along with the Tohoku region, Hokuriku is also known as a famous sake producing region. The sake produced in Niigata and Toyama prefectures features a light flavor and Ishikawa and Fukui prefectures feature a rich and strong flavor.

KANTO
Sake produced in the Kanto region often features a light flavor.

CHUBU
Sake produced in Shizuoka prefecture is light and mild while in Aichi and Gifu prefectures it is a rich and strong flavor.

KINKI
The sake features a rich and dry flavor. Yamadanishiki rice, now recognized worldwide as "The King of Brewing Rice", is also produced in Hyogo. This region's development in sake brewing is blessed with high quality water as well.

CHUGOKU
Hiroshima prefecture in particular has emerged as a sake producing area since the later Meiji era. The sake features a soft and smooth flavor.

SHIKOKU
The sake produced in Kochi prefecture features a crisp light flavor while the sake produced in Ehime prefecture is similar to Hiroshima's with a soft and smooth flavor.

KYUSHU
This region is well-known as a shochu producing area and famous for its sake breweries. The sake features a sweet flavor while maintaining depth.

第四章

买日本酒：
那些品位独到的酒水臻选店

PURCHASE:
SAKE SHOPS WITH
UNIQUE SELECTIONS

在品质管理卓越的酒屋，找寻属于自己的那瓶真爱之酒

田端酒造位于和歌山县内，是始创于1851年（嘉永四年）的老字号酒藏，长谷川聪子是它的第七代传人。据说在过去酒藏是禁止女性出入的。"不过我母亲继承了家业，成了第六代传人，所以我感觉这些年藏元也在一点点地发生着变化。"长谷川小姐说道。她说自己学生时代没什么钱时曾被祖父（田端酒造的第五代传人）教育说："别在外面喝那些劣质、便宜的酒，要喝就在家里喝。"长谷川现在定居东京，因工作原因经常在和歌山和东京间往返。对于选酒，她认为："一家店酒品是否丰富固然重要，但更重要的是品质管理有没有做好。毕竟酒是活的，有生命的。"虽然这次没能具体介绍，不过长谷川也非常推荐去百货商店的酒类卖场选购，因为那里不仅可以跟店员咨询，还可以试饮。

第四章

Find your favorite brand at liquor stores committed to quality control

Satoko Hasegawa is a seventh-generation sake brewer at Tabata Sake Brewery established in 1851 in Wakayama. "Women were not allowed to enter any sake warehouses before, but since my mother succeeded a sixth-generation sake brewer, breweries' methods have been changed." she explained. Satoko was told during her days as a poor student "don't drink bad sake, don't drink cheap sake, drink sake at home" by her grandfather and also a fifth-generation brewer. She currently lives in Tokyo but often travels between Wakayama and Tokyo. "The key point for choosing sake in liquor stores is the selection size but also the level of quality control because sake is alive." she advised. Although it's not covered in this edition, she also recommends shops at department stores because you can ask questions and try tasting the sake.

ARTRIP ADVISER
艺术之旅顾问

长谷川聪子
Satoko Hasegawa

田端酒造第七代传人。其祖父，同时也是酒造第五代传人，酿造的"罗生门"清酒享誉日本。她在继承传统的同时锐意创新，研发了自己的原创酒——"聪子的酒"，是颇受关注的青年清酒酿酒师。

"Rashomon" created by her grandfather and a fifth-generation brewer is widely recognized as a premium sake. While keeping the tradition, Satoko is in the spotlight recently as a young sake brewer who produces her original brand "Satoko no sake."

❶⁶ 藏家（町田）

内行人都知道的超人气酒类专卖店。即使藏家附近的交通不算方便，照样吸引了络绎不绝的顾客驾车或步行前来。藏家由现任店主浅沼芳征的父亲创立，浅沼芳征则是第二代。面积不大的店内摆满了各种酒。清酒和烧酒在一层，葡萄酒在二层，另有供试饮或举办活动的场地。店内约有不到一百种清酒，浅沼说他们在挑选清酒时把每次选酒当作与酒的一次邂逅，不会受到酒藏规模或时下清酒流行趋势的影响。"就算是那种只有当地人才知道的小作坊，只要有好酒，我们就会锲而不舍地请求对方允许我们进货销售。"进货后，浅沼会像对待自己的孩子一样悉心照料并保存每一款酒，一丝不苟地做好温度和品质管理，精心设计店内的海报，直至将酒送到顾客面前。浅沼认为酒类专卖店的使命就是"将酿酒师的理念完整地传递给顾客，让顾客在饮用时可以发自内心地觉得好喝"。也正因为此，浅沼始终坚持面对面服务顾客，将酒的品质与特点传达给客人。

⓽ 星期二至星期六 9:30—20:00，星期日 9:30—19:00 ⓭ 042-793-2176 ⓱ 星期一 ⓰ 町田市木曽西 1-1-15 ⓧ 在 JR 或小田急线町田站乘巴士 10 分钟 ⓦ kura-ya.com

⓰ **KURAYA**（Machida）

Even though Kuraya is located far from either Machida or Kobuchi station, this local liquor store bustles with customers arriving by car or on foot. Established by the previous owner and then succeeded by a second-generation owner of Yoshimasa Asanuma, Kuraya has a large selection of sake and shochu on the 1st floor and wine on the 2nd floor, as well as a space for tasting and events. Yoshimasa selects almost 100 kinds of sake, "rather than the size of a brewery or market trends, even if a brewery is only locally-known and small but produces excellent sake, I would tenaciously negotiate with them to have their brand in my store because we treasure every single encounter. We take good care of our products as if our own children, and commit to good quality including temperature control. Our goal as a liquor store is to deliver the producer's message to our customers and to wish that they truly enjoy the flavor when they drink. This is the reason why we really care about a face-to-face sales approach."

(H) Tue-Sat 9:30-20:00, Sun 9:30-19:00　(T) 042-793-2176　(C) Mon　(Ad) 1-1-15 Kisonishi, Machida
(Ac) 10-min. by bus from Machida Station (JR Line & Odakyu Line)　(U) kura-ya.com

---

KURAYA

悬挂在天花板的手绘招牌可以帮助客人找到清酒、烧酒等各种酒所在的位置。据说在店铺刚刚创立的时候是由浅沼的父亲负责管理并销售葡萄酒,母亲则负责清酒事宜。

Handmade POP signage hanging from the ceiling makes it easy for customers to find where the products such as sake and shochu are placed. When Yoshimasa's father ran the business, he was in charge of wine and his mother was in charge of sake.

包含风干食品、罐头在内的佐酒零食,琳琅满目地摆满货架。收银台旁边还有专门的罐头柜台,让人不知不觉就买了好多别处少见的稀罕物。每件商品上都贴着店铺手写的介绍卡片,实在让人惊喜。

There are plenty of appetizers which accompany sake such as dry and tinned snacks. The customers probably end up purchasing those unique tinned items which are displayed next to the cash register counter. Looking at the recommendations on POP signs is fun to do.

浅沼的母亲通过个人的努力获得了名酒"罗生门"的代理权,这款和歌山地方酒一般很少出现在酒类专卖店。4年前从大型广告公司辞职回家接手老店经营的浅沼说:"虽然父亲已经去世,但母亲还健在,我还有许多东西要跟她学习。"

With the owner's mother's effort, Kuraya now carries local sake "Rashomon" which is not available at many stores. "My father passed away but my mother is in good health. I still learn a lot from her." says Yoshimasa who resigned from a major advertising agency and became the owner 4 years ago.

花泉酒造位于福岛县会津地区。该酒造出产的酒往往有市无价,市面上极难买到。藏家是它的特许经销商。为了让藏家成为"深受顾客及藏元信赖的专卖店",浅沼一有时间就去各地酒造拜访。

Kuyara has an exclusive dealer contract with Hanaizumi Sake Brewery which produces the local sake in the Aizu region of Fukushima prefecture. Their brands are rarely offered on the market. The owner explores as many local breweries as possible to build their credibility among both customers and breweries.

第四章

近来在清酒市场颇受关注的古酒。长期熟成让酒的口感变得更加柔和,其独特风味俘获了不少酒客的心。"我们只卖自己真心觉得好喝的酒。"浅沼介绍道。

Koshu recently emerged in the sake market. Sake develops a smoother flavor over time through the long aging process and this unique flavor attracts many sake fans. Yoshimasa says, "we only carry brands which I really savor myself."

藏家店内还可以购买到用酒糟做成的点心和甜酒,以及使用清酒制作的化妆品。客人也可以去开在町田站附近的姊妹店居酒屋"藏家SAKERABO"试试,那里酒水供应丰富,设有立席,最小点单量仅为"半半合"。

The confectioneries made with sake lees, amazake and the cosmetics containing sake are available at the store. Opened in front of Machida station as a satellite shop, sake bar KURAYA SAKERABO serves sake with a minimum size of 45 ml to enjoy different varieties.

KURAYA

⑰ 小山商店（多摩）

小山商店位于多摩市，1914年（大正三年）开业，店内酒品之丰富被誉为全日本第一，众多酒客慕名而来。第三代店主小山喜八不迷信名牌酒造，常亲自拜访藏元，找寻味美价廉及饱含酿酒师心意的还不为人所熟知的好酒。店内酒品约有八成来自酒厂直销。店内货架上常年陈列着五百余种清酒，规模壮观，堪称"清酒图书馆"。小山商店的宗旨是"将酿造人的理念准确地传递给顾客"，故每一瓶酒都经过精挑细选，店主可以自信满满地推荐给客人。所有酒都被精心保存在约80平方米的冷藏库和即使在盛夏时节温度也不会高于9度的地下酒窖中。店主集结当地同好组成学习小组——"多摩独酌会"；每两月举行一次活动，在活动中店主会邀请酒藏工作人员和清酒爱好者们共同学习研究各款清酒的历史，探究各款清酒与料理的搭配，发掘尚不为人所知的美味清酒。店内除清酒外还有种类丰富的烧酒和日本本土有机葡萄酒。

(时) 星期一至星期六 9:00—20:00，星期日 10:00—19:00 (电) 042-375-7026 (休) 每月第三个星期日
(址) 多摩市关户 5-15-17 (交) 京王线圣迹樱丘站步行 13 分钟 (网) www.sake180.cc

⑰ KOYAMASHOTEN (Tama)

Established in 1914 in Tama city, Koyamashoten is acclaimed as a store with the largest variety of local sake in Japan. Many patrons visit from far and wide. The third-generation owner Kihachi Koyama often visits breweries to look for a truly delicious sake. His selection is mainly reasonably-priced tasty sake that is made with the brewery's passion, so 80 percent of items come directly from the breweries. 500 sake are regularly placed on the shelves like a library. Each and every bottle is stored in about 83 ㎡ of fridge space and a cellar that is kept below 9 °C. With their motto "commitment to deliver producer's message to the customers", they promote every item with confidence. A study group, Tamadokushakukai was formed among some store owners to gain a deep understanding of sake and discover unknown delicious sake through the event which is held every two months with guest breweries. Shochu and domestic organic wine are also available at the store.

(H) 9:00-20:00, Sun 10:00-19:00 (T) 042-375-7026 (C) Third Sunday of the month (Ad) 5-15-17 Sekido, Tama (Ac) 13-min. walk from Seisekisakuragaoka Station (Keio Line) (U) www.sake180.cc

KOYAMASHOTEN

⑱ 升本（虎之门）

始创于 1940 年的老字号酒类专卖店，位于政府机关聚集之地的虎之门。过去来店里光顾的一般是附近工作的上班族，近来外国客人愈发增多。店内经销 600 种以上酒饮；其中清酒品种最为繁多，从价格低廉的小瓶装到一升大瓶装，再到如工艺品般精美的酒瓶装，可以满足顾客的不同需求。这些年来，升本一直在紧跟时代需求，积极开展多元经营。除本店隔壁的葡萄酒专卖店"Vin sur Vin"、马路对面的居酒屋"升本"外，"升本"作为品牌公司还涉足葡萄酒进口贸易，并且正与知名酒造携手计划将清酒推向世界。对于自家的选酒方针，店主篠原义昌觉得不需要考虑销量，不用追逐潮流，坚持要选"那些可以感受到酿酒师心意、知根知底值得信赖的酒"。此外，店内排列整齐的酒瓶、微凉的室内温度、一尘不染的地板，这些细节无一不体现着店主对于酒的品质管理的用心。

(时) 10:00—19:30 (电) 03-3501-2810 (休) 星期六、星期日、节假日 (址) 港区虎之门 1-7-6 (交) 地铁虎之门站步行 1 分钟 (网) 无

⑱ **MASUMOTO** (Toranomon)

Surrounded by government offices in Toranomon, Masumoto was established in 1940 and has a long rich history. Masumoto has been popular among the office workers but also among foreign customers recently. Over 600 popular sake are regularly available and there are a variety of sizes from a cheaper one, a 1800 ml bottle one, to the limited handcrafted bottle to satisfy customer's needs. Masumoto cooperates with some breweries to import wine but also export sake in order to promote sake widely around the world. The management diversifies the business to open Wine Boutique Vin sur Vin and an izakaya Masumoto to meet the current demands. The owner, Yoshimasa Shinohara says, "Masumoto's policy is selecting brands reflecting the producer's passion so that customers can see their message, rather than selling popular or trendy items." Neatly arranged bottles, a nicely chilled interior and a clean floor without even a speck of dust are strong indicators of their high level of commitment to quality control.

(H)10:00-19:30 (T)03-3501-2810 (C)Sat, Sun & National holidays (Ad)1-7-6 Toranomon, Minato-ku (Ac)1-min. walk from Toranomon Station (Tokyo Metro Ginza Line) (U)N/A

MASUMOTO

靠着上代及上上代与酒商结下的深厚关系，在升本可以买到像"罗生门"、"獭祭"这种平时很难入手的名酒。店主娘：我们有许多市面上不常见的酒，感兴趣的话欢迎询问店员。

With a solid relationship between previous owners and breweries, "Rashomon" and "Dassai" rarely offered on the market are available at Masumoto. "Some uncommon brands among the liquor stores may be found here so ask the staff about them."

在老字号酒类专卖店升本的直营餐厅里喝上一杯清酒
Savor sake at Masumoto's long-established outlet

虎之门 升本（虎之门）

1969 年开业。因价格经济实惠，在周边工作的上班族间人气很高，许多人选择下班后来喝上一杯再回家。近来年轻女性及外国客人明显增多。每日供应含热行酒在内的 30 余种清酒。

(时) 16:30—22:00　(电) 03-3591-1606　(休) 星期六、星期日、节假日　(址) 港区虎之门 1-18-16　(交) 地铁虎之门站步行 1 分钟　(网) 无

Toranomon MASUMOTO（Toranomon）

Established in 1969 and serving sake at a reasonable price, Toranomon Masumoto has been popular among the businessmen who work nearby the store, but is recently bustling with young, female and foreign customers. Around 30 kinds of sake are usually available including popular items.

(H) 16:30-22:00　(T) 03-3591-1606　(C) Sat, Sun & National holidays　(Ad) 1-18-16 Toranomon, Minato-ku　(Ac) 1-min. walk from Toranomon Station (Tokyo Metro Ginza Line)　(U) N/A

第四章

从刺身、烧烤到使用当季食材烹制的下酒菜，菜单品种丰富。每卖完一道菜就会从墙上摘去相应菜牌。

Octopus oden, sashimi, grilled foods and seasonable appetizers are popular items on the menu among the patrons. Once they are sold out, the tag is removed from the menu board.

店铺位于大马路旁的小胡同里，标志是门前的这张暖帘。1～3 层均有座位。3 层有楼层限定的特殊套餐（需预约），可自带酒水。

Look for this curtain to find Masumoto tucked away on the side street. At this three-storey bar, you can bring your own sake and enjoy the special menus on the 3rd floor (advance booking is required).

⓳ 内藤商店（五反田）

距东急目黑线不动前站不远的内藤商店，据说在 1923 年（大正十二年）创业时原本是一间酒藏，后慢慢发展成酒屋。店内目前由第四代店主东条晃一及其父亲运营，主要贩卖适合日常饮用的味美质优的酒饮。"我们的目标是成为一家植根本地、为周边老百姓服务的酒类专卖店。清酒也有潮流和变化，有的酒也许一时卖得不错，但客人不会反复购买。"所以，内藤商店要卖的是"味美价廉、让人喝了一次还想再喝的酒"。近些年来，越来越多的年轻酿酒师开始崭露头角，东条说："现在只要一有时间，我就去拜访各地酒造，争取多走几家，支持下年轻的酿酒人。"店内一个摆满了玻璃易开杯的角落人气很高。内藤商店同时也是著名的烧酒商店，烧酒品种丰富。店内 900 余种一升装的大酒瓶排列整齐，壮观非凡。

(时) 9:30—21:30　(电) 03-3493-6565　(休) 星期日、节假日　(址) 品川区西五反田 5-3-9　(交) JR 五反田站步行 8 分钟　(网) www.facebook.com/naitohshoten

⓴ **NAITOSHOTEN** (Gotanda)

Located nearby Fudomae station, Naitoshoten was originally established as a brewery in 1923. The fourth-generation owner Koichi Tojo now operates this liquor shop with his father Tatsuo. Their motto is to provide delicious and good quality sake that customers can enjoy daily. "Our goal is to become a specialty store rooted in the local area. Since there are industry trends, we occasionally sell certain brands for a limited time period, but it's not what customers want to drink again. That is why we select the brands that are reasonably-priced and easy to find. I make efforts to visit as many breweries as I can to support the increasing number of young brewers." At the store, there is a special area featuring one cup-sized sake and a large selection of shochu. The sight of their regular stock of 900 kinds of 1800 ml-sized sake bottles is astounding.

(H) 9:30-21:30 (T) 03-3493-6565 (C) Sun & National holidays (Ad) 5-3-9 Nishigotanda, Shinagawa-ku (Ac) 8-min. walk from Gotanda Station (JR Line) (U) www.facebook.com/naitohshoten

内藤商店

内藤商店出售的每一瓶酒都经过严格的品质甄选。在这里还可以买到内藤商店与酒藏的联名限定商品。"酒是酿酒师心血的结晶。将酿酒师的心血与对酒的热忱之情恭恭敬敬地传递给客人是我们的职责。"

Each and every bottle is kept under superb quality control. Limited editions are created through collaboration between Naitoshoten and breweries. "Sake reflects the passion of a brewery. As a liquor store, we deliver their messages with respect to the customers."

商店深处有一角堆满了日本各地地方酒的玻璃易开杯。易开杯无论容量还是价格都比一升瓶更亲民，可以多试几种酒，很受外国顾客欢迎。

At the back of the store, there is a special area of one cup-sized local sake from all over Japan. Unlike 1800 ml-sized sake, a cheaper price range and wide variety of one cup-sized sake seems to attract many customers, especially foreigners.

NAITOSHOTEN

❷⓿ **STAND BAR MARU**（八丁堀）

拥有三十多年历史的立饮屋 STAND BAR MARU 位于紧邻八丁堀站出口的写字楼群中一个不起眼的小角落。所谓立饮，顾名思义就是"站着喝"。STAND BAR MARU 是现在遍地开花的立饮式居酒屋的始祖。日本以前就有在酒类专卖店买了酒后直接站在店内一角喝完的文化，日语叫"角打"，不过立饮居酒屋形态的确立却是近些年的事情。店名寓意"大家通过酒联结在一起"，是店铺重装时设计师的创意。MARU 由酒类专卖店宫田屋运营，吧台和餐桌就设在宫田屋的旁边。一层立席主要提供清酒和啤酒；二层和三层设有座席，提供葡萄酒等酒饮。菜单品种丰富，价格实惠。"很多客人都是自己一个人来，简单喝上一杯就走；当然也有结伴来的，一群人热热闹闹。无论哪种我们都欢迎，真诚期待大家的光临。"

(时) 16:30—23:00 (电) 无 (休) 不固定 (址) 中央区八丁堀 3-22-10 (交) 地铁八丁堀站步行 1 分钟
(网) maru-miyataya.com

❷⓿ **STAND BAR MARU** (Hacchobori)

Located nearby Hacchobori station and surrounded by office buildings, Stand Bar Maru has a 30-year history as a standing-only bar. It is also recognized as one of the first standing-only bars in the industry. In the old days, there were traditional liquor stores called "kakuuchi" where you could sample various sake with nibbles. It's been a few years since the current style of standing-only bars emerged. A message of "creating a circle of people through sake" is embedded in the name of Maru, which is suggested by the designer involved in the current building's renovation. Operated by a liquor store Miyataya and adjoined to the store, Maru has a counter and tables. Customers enjoy sake and beer on the 1st floor, and wine on the 2nd and 3rd floors in a relaxed seated area. A variety of Kyoto style home-cooked appetizers are popular as well. "Many customers come alone but also groups of customers come with their companions. We want many people to drop by and enjoy drinks at our bar."

(H) 16:30-23:00 (T) N/A (C) Irregular (Ad) 3-22-10 Hacchobori, Chuo-ku (Ac) 1-min. walk from Hacchobori Station (Tokyo Metro Hibiya Line) (U) maru-miyataya.com

㉑ 合羽桥 酒之SANWA（合羽桥）

1955年，"酒之SANWA"店铺在上野开业。2015年，其二号店在合羽桥道具街开业。合羽桥店只贩卖720毫升装的清酒，提供试饮。酒之SANWA在选酒时不重酒品的名气，店内清酒大多来自品质优良的小型造造，主要合作的品牌有：上喜元、阿部勘、乾坤一、凤凰美田、岩、浅间山、东魁盛、白隐正宗、十九、龟之海、小左卫门、白岳仙、三重锦、花巴、奥播磨、美和樱、宝剑、开春、悦凯阵、七田等。店内吧台常年供应着数十种清酒，花上500日元就可以试饮两种纯米酒、纯米吟酿，或一种大吟酿。当然如果喜欢的话，也可以直接在店内购买，每瓶720毫升，价格都十分亲民。许多对清酒感兴趣但由于了解不多不知从何下手的酒友都可以来这家店一试。店内除清酒外，还贩卖本格烧酒及产自葡萄牙马德拉群岛的马德拉葡萄酒。

(时) 12:00—18:00 (电) 03-5830-3521 (休) 不固定 (址) 台东区松之谷 3-17-11 (交) 地铁田原町站步行 11 分钟 (网) sanwasake.jp

㉑ KAPPABASHI SANWA 720（Kappabashi）

Opened in 2015 in Kappabashi Dougugai as a second outlet of Sakenosanwa which was founded in 1955 in Ueno, KAPPABASHI SANWA 720 is a specialty store which carries only 720 ml-sized sake. Customers can enjoy sake tasting at the store. Their selection is not particular about major brands but focused on high-quality sake and the brewery size does not matter. Customers can find brands including Jokigen, Abekan, Kenkonichi, Hououbiden, Iwao, Asamayama, Tokaizakari, Hakuinmasamune, Juku, Kamenoumi, Kozaemon, Hakugakusen, Mienishiki, Hanatomoe, Okuharima, Miwasakura, Hoken, Kaishun, Yorokobi-Gaijin and Shichida. 10 kinds of seasonable sake are regularly offered for tasting. You can taste 2 kinds of Junmai or Junmai Ginjo, or one kind of Dai Ginjo for 500 yen and then purchase a full-sized bottle at the store if you wish. All are at a reasonable price in a 720 ml size. This tasting has a good reputation among customers who want to try but hesitate to buy without knowing the flavor. In addition to sake, authentic shochu and Madeira wine produced in Madeira Island, Portugal are available at this store.

(H) 12:00-18:00　(T) 03-5830-3521　(C) Irregular　(Ad) 3-17-11 Matsugaya, Taito-ku　(Ac) 11-min. walk from Tawaramachi Station (Tokyo Metro Ginza Line)　(U) sanwasake.jp

KAPPABASHI SANWA 720

㉒ KURAND SAKE MARKET(池袋)

100 种清酒,不限时畅饮,允许自带料理及下酒菜,每人收费 3000 日元。这就是这家店人气高的秘密。设立了这一独特制度的清酒专门店 KURAND SAKE MARKET 从 2015 年在池袋开业以来,已在东京都内开了 8 家分店。为了让客人品尝到"好喝但不出名"或"有故事"的清酒,店员遍访全国酒造,严选真正美味好酒。店内另有适合不擅长饮酒的人士饮用的原创气泡清酒和低度清酒。客人可按个人喜好自行从冰箱取酒,酒器也可以自由选择。有的客人会从百货商店地下食品卖场打包下酒菜带到店内享用,有的客人更出其不意一些,会自带中华料理或民族特色食品。店里不时会提供一些创意清酒饮法,比如夏天用碳酸水兑大吟酿调成"大吟酿嗨棒",冬天在温热的清酒上扣一勺冰激凌制成"清酒阿芙佳朵"。

(时) 星期一至星期五 17:00—23:00(最后点单时间 22:45),星期六 12:00—16:00(最后点单时间 15:30),星期日 17:00—23:00(最后点单时间 22:45) (电) 03-6912-6686 (休) 无 (址) 丰岛区西池袋 3-27-3 S&K 大厦 4 层 (交) JR 池袋站步行 3 分钟 (网) kurand.jp/sakemarket/ikebukuro

㉒ KURAND SAKE MARKET (Ikebukuro)

This sake specialty store offers a unique style of service. Customers can enjoy 100 kinds of sake for an unlimited time and are free to bring foods and nibbles (¥ 3,000 per person). After the first outlet was opened in Ikebukuro in 2015, 8 outlets are now open in the Tokyo area (as of September 2018) .With their motto of promoting sake which have yet to enter the spotlight but have their own unique stories and messages, they visit breweries all over Japan to find truly delicious sake. For sake beginners, Kurand original sparkling sake and low-alcohol sake are available. You can take your favorite brand of sake from the fridge and choose your favorite sake cup. Some customers bring take-out foods from department store basement food halls, while others bring surprising choices such as ethnic or Chinese foods. Other than normal ways of drinking sake, they create "out-of-the box" innovative menus such as "Dai Ginjo highball", which is a mixture of Dai Ginjo and soda in summer, and "sake affogato", which involves putting hot sake on top of ice cream in winter.

(H) Mon-Fri 17:00-23:00 (L.C.22:45), Sat & Sun 12:00-16:00 (L.C.15:30), 17:00-23:00 (L.C.22:45)
(T) 03-6912-6686 (C) None (Ad) 4th floor S&K Bldg, 3-27-3 Nishiikebukuro, Toshima-ku (Ac) 3-min. walk from Ikebukuro Station (JR Line) (U) kurand.jp/sakemarket/ikebukuro

在日本各地的乡土馆入手地方清酒
PURCHASING SAKE AT LOCAL SPECIALTY STORES

有乐町的东京交通会馆称得上是乡土馆汇聚地,其中就有和歌山县开设的"和歌山纪州馆",于2004年开业,主要贩卖和歌山县特产。馆内商品种类丰富,每年有800余种商品在此登场,颇受顾客喜爱。和歌山县的特产是橘子和梅干,其实和歌山出产的清酒也不错。流经奈良的纪之川沿岸分布了众多酒藏,酿出的地方酒中有不少都很出名。就着馆内贩卖的农产品、风干食品,品尝来自同一片土地的清酒,想必别有一番滋味。

Opened in 2004 at Tokyo Transportation Hall in Yurakucho which is known as "the mecca of local specialty shops," Wakayama Kishukan sells local products from Wakayama prefecture. They carry over 800 kinds of products every year. A large selection of items attracts many customers. Wakayama is well known for oranges and *umeboshi* (pickled Japanese plum), but also premium local sake which are brewed by a number of breweries located along the Kinokawa river that runs through Nara prefecture. Drinking the delicious local sake accompanied by locally-produced farm products and dried nibbles is certainly a unique experience.

❷❸ 和歌山纪州馆(银座)

(时)星期一至星期六 10:00—19:00,星期日、节假日 10:00—18:00 (电)03-6269-9434 (休)无(年末年初除外) (址)千代田区有乐町 2-10-1 东京交通会馆地下 1 层 (交)JR 有乐町站步行 1 分钟 (网)www.kishukan.com

❷❸ WAKAYAMA KISHUKAN
(Ginza)

(H) 10:00-19:00, Sun & National holidays 10:00-18:00 (T) 03-6269-9434 (C) Non (Except for Year-end and New Year holidays) (Ad) B1F Tokyo Transportation Hall, 2-10-1 Yurakucho, Chiyoda-ku (Ac) 1-min. walk from Yurakucho Station (JR Line) (U) www.kishukan.com

和歌山是日本的水果盛产地,除橘子外还盛产梅子、桃子、草莓、柿子和贾巴拉柑橘,一年四季都有新鲜水果上市。和歌山同时也是旅游胜地,县内有高野山和熊野三山等多处世界遗产。

Wakayama prefecture is known as Japan's leading producer of fruits year-round such as oranges, plums, peaches, strawberries, persimmons and jabara (a type of Japanese citrus). There are also famous tourist attractions including the World Heritage Sites of Koyasan and Kumanosanzan.

店内可以买到包括田端酒造的名酒罗生门在内的众多人气清酒。从小小的易开杯到四合瓶,容量选择丰富。作为著名的梅子产地的和歌山,自然也盛产梅酒。用烧酒、清酒或白兰地做基酒调制的 20 余种梅酒中,一定有符合你口味的那一款。

Their selection has a variety of sake from one-cup to 4-go (720 ml) sized bottles including a popular Rashomon brand of Tabata Sake Brewery. As a leading plum producing region, 20 kinds of ume-shu are sold at the store. The base spirits vary from shochu to sake and brandy. You can find your favorite ume-shu.

乡土馆名录
LIST OF LOCAL SPECIALTY SHOPS

岩手 银河广场（银座）

位于银座歌舞伎座对面，店内面积宽敞。
除清酒外还有许多其他种类丰富的岩手县特产。

(时) 10:30—19:00（每月最后一天营业至 17:00）(电) 03-3524-8282 (休) 无（年末年初除外）(址) 中央区银座 5-15-1 南海东京大厦 1 层 (交) 地铁或都营地铁浅草线东银座站步行 1 分钟 (网) www.iwate-ginpla.net/index.html

IWATE GINGA PLAZA (Ginza)

Located in front of Kabukiza. Featuring specialties from Iwate.

(H) 10:30-19:00 (last day of the month 10:30-17:00) (T) 03-3524-8282 (C) None (Except for Year-end and New Year holidays) (Ad) 1st floor Nankaitokyo Bldg. 5-15-1 Ginza Chuo-ku (Ac) 1-min. walk from Higashiginza Station (Tokyo Metro Asakusa Line) (U) www.iwate-ginpla.net/index.html

日本桥福岛馆-MIDETTE（日本桥）

2014 年开业的福岛县特产直销商店。
内设清酒试饮柜台。

(时) 星期一至星期五 10:30—20:00，星期六、星期日及节假日 11:00—18:00 (电) 03-6262-3977 (休) 年末年初 (址) 中央区日本桥室町 4-3-16 柳屋太洋大厦 1 层 (交) 地铁三越前站步行 3 分钟 (网) midette.com

NIHONBASHI FUKUSHIMAKAN — MIDETTE (Nihonbashi)

Opened as Fukushima's antenna shop with a sake-tasting counter.

(H) 10:30-20:00, Sat, Sun & National holidays 11:00-18:00 (T) 03-6262-3977 (C) Year-end and New Year holidays (Ad) 1st floor Yanagiyatayo Bldg. 4-3-16 Muromachi Nihonbashi Chuo-ku (Ac) 3-min. walk from Mitsukoshimae Station (Tokyo Metro Ginza Line) (U) midette.com

日本桥富山馆（日本桥）

名酒之乡富山县的特产直销商店。另有一间分店在有乐町。
内部空间可举行活动，另设有可供品鉴清酒的酒吧。

(时) 10:30—19:30（商店）(电) 03-6262-2723 (休) 年末年初 (址) 中央区日本桥室町 1-2-6 日本桥大荣大厦 1 层 (交) 地铁三越前站步行 1 分钟 (网) toyamakan.jp

NIHONBASHI TOYAMAKAN (Nihonbashi)

Toyama's antenna shop. Yurakucho branch. A bar offering sake sampling.

(H) 10:30-19:30(shop) (T) 03-6262-2723 (C) Year-end and New Year holidays (Ad) 1st floor Nihonbashi Taiei Bldg. 1-2-6 Muromachi Nihonbashi, Chuo-ku (Ac) 1-min. walk from Mitsukoshimae Station (Tokyo Metro Ginza Line) (U) toyamakan.jp

乡土馆名录
LIST OF LOCAL SPECIALTY SHOPS

表参道 新潟馆 N'ESPACE（表参道）

店内地方酒包含约 90 家酒藏的 300 余种品牌，其中有些只在当地贩售。
附设餐厅及观光·新潟就职信息中心。

(时) 10:30—19:30 (电) 03-5771-7711 (休) 无 (址) 涩谷区神宫前 4-11-7 (交) 地铁表参道站步行 1 分钟 (网) www.nico.or.jp/nespace

OMOTESANDO NIIGATAKAN N'ESPACE (Omotesando)

Featuring breweries and sake from Niigata. Various information offered.

(H) 10:30-19:30 (T) 03-5771-7711 (C) None (Ad) 4-11-7 Jingumae, Shibuya-ku (Ac) 1-min. walk from Omotesando Station (Tokyo Metro Ginza Line) (U) www.nico.or.jp/nespace

广岛品牌店 TAU（银座）

三层建筑。二层设有广岛地方酒专门柜台，提供试饮。
另设有广岛烧专门店等餐饮店。

(时) 10:30—20:00 (电) 03-5579-9952（日本酒工作室 翠）(休) 无 (址) 中央区银座 1-6-10 银座上一大厦 (交) 地铁银座一丁目站步行 1 分钟 (网) www.tau-hiroshima.jp

HIROSIMA BRANDSHOP TAU (Ginza)

Featuring local sake from Hiroshima. Various outlets including Okonomiyaki shop.

(H) 10:30-20:00 (T) 03-5579-9952 (Sake kobo Midori) (C) None (Ad) Ginza Kamiichi Bldg. 1-6-10 Ginza, Chuo-ku (Ac) 1-min. walk from Ginza 1-chome Station (Tokyo Metro Ginza Line) (U) www.tau-hiroshima.jp

MARUGOTO 高知（银座）

店内可以买到"醉鲸""龟泉"等高知县名酒。在二层餐厅一边享受
高知食材烹制的料理，一边品味清酒也是个不错的选择。

(时) 10:30—20:00 (电) 03-3538-4365 (休) 无 (址) 中央区银座 1-3-13 (交) JR 有乐町站步行 3 分钟 (网) www.marugotokochi.com

MARUGOTO KOCHI (Ginza)

Featuring "Suigei" and "Kameizumi" from Kochi. Restaurant located on 2nd floor.

(H) 10:30-20:00 (T) 03-3538-4365 (C) None (Ad) 1-3-13 Ginza Chuo-ku (Ac) 3-min. walk from Yurakucho Station (JR Line) (U) www.marugotokochi.com

第四章

小百科 / GLOSSARY

酒藏：酿酒厂，也指代酒窖。

酒造：与"酒藏"类似，指酿酒厂。

藏元：酒厂经营者。

合：日本酒的计量单位。一合约 180 毫升。"半合"约 90 毫升，"半半合"约 45 毫升。

烧酒：日本指用薯类和谷物制造的蒸馏酒，酒精含量在 20%～40%。

泡盛：一种特产于日本冲绳县的蒸馏酒，做法十分讲究，属于烧酒类。酒精度数有 50 度、60 度、120 度。熟成三年及三年以上的泡盛被称为"古酒"。

甜料酒：日语写作"味淋"，指在烧酒和糯米里掺入酒曲做成的又甜又浓的酒，用作调料。

开镜：为樽酒开桶的仪式。酒樽上盖为"镜"，象征圆满，砸开后上盖四裂象征逐渐繁荣。

《日本书纪》：日本最早的敕撰史书，记述自神代至持统天皇时代的日本正史。

精米步合度：指酒米米粒被磨去外层后所剩部分占原米的比例，是影响清酒口味的重要因素之一。

德利：细长的口部狭小的盛酒器皿。

猪口：小酒杯，主要指陶瓷器或银、锡制的小酒杯。

辛口：味道辛辣的酒。

甘口：带甜味的酒。

嗨棒：音译自英文"Highball"，指以烈酒为基础加入碳酸水调制的酒。

NOTES

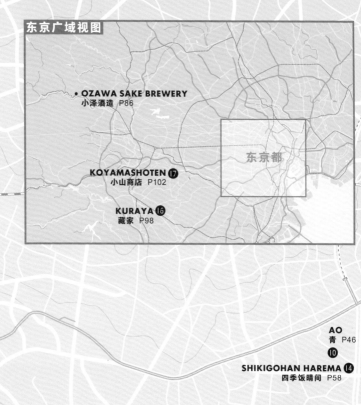

TOKYO ARTRIP SAKE by BIJUTSU SHUPPAN-SHA TOKYO ARTRIP Editorial Team
Copyright © BIJUTSU SHUPPAN-SHA TOKYO ARTRIP Editorial Team, ® Bijutsu Shuppansha
All rights reserved.
Original Japanese edition published by Bijutsu Shuppan-Sha Co., Ltd., Tokyo.
This Simplified Chinese language edition is published by arrangement with Bijutsu
Shuppan-Sha Co., Ltd., Tokyo in care of CITIC PRESS JAPAN CO., LTD, Tokyo

本书仅限中国大陆地区发行销售

日文版工作人员

Cover Illustration: NORITAKE　　Designer: TUESDAY (Tomohiro+Chiyo Togawa)
Map: Manami Yamamoto (DIG.Factory)
Photographer: Takashi Matsumura（except p34~35, 38~39, 40~41, 54~55, 58~59, 66, 71~81, 86~91, 102~103,114~115, 116~117, 119）
Japanese Writer: Hana Hasegawa　　Translator: Akiko Ebihara-Cleaver
Proofreader (Japanese): Mine Kobo　　Proofreader (English): Jonathan Berry
Editorial Director: Miki Usui (BIJUTSU SHUPPAN-SHA CO., LTD.)

图书在版编目（CIP）数据

日本酒 / 日本美术出版社书籍编辑部编著；朴惠译
. -- 北京：中信出版社，2019.7（2022.3 重印）
（东京艺术之旅）
ISBN 978-7-5217-0418-1

Ⅰ. ①日… Ⅱ. ①日… ②朴… Ⅲ. ①酒文化 - 介绍
- 日本 Ⅳ. ① TS971.22

中国版本图书馆 CIP 数据核字 (2019) 第 073280 号

日本酒

编　　著：【日】美术出版社书籍编辑部
译　　者：朴惠
出版发行：中信出版集团股份有限公司
　　　　　（北京市朝阳区惠新东街甲4号富盛大厦2座　邮编　100029）
承　印　者：北京雅昌艺术印刷有限公司

开　　本：880mm×1230mm　1/32　　印　张：4
字　　数：68千字　　　　　　　　　版　次：2019年7月第1版
印　　次：2022年3月第5次印刷　　　京权图字：01-2019-2347
广告经营许可证：京朝工商广字第8087号
书　　号：ISBN 978-7-5217-0418-1
定　　价：45.00元

版权所有·侵权必究
如有印刷、装订问题，本公司负责调换。
服务热线：400-600-8099
投稿邮箱：author@citicpub.com